Astronomy for Young and Old

ASTRONOMY
FOR YOUNG AND OLD

A BEGINNER'S GUIDE TO
THE VISIBLE SKY

Walter Kraul

Floris Books

Translated and edited by Christian Maclean
Illustrated by Dazze Kamerl

First published in German under the title *Erscheinungen am Sternenhimmel*
by Verlag Freies Geistesleben in 2002
First published in English by Floris Books in 2014
© 2002 Verlag Freies Geistesleben & Urachhaus GmbH
English version © 2014 Floris Books

British Library CIP Data available
ISBN 978-178250-046-9
Printed in Poland

Contents

Foreword

If people stop looking up at the stars, they will fall down.
Chinese proverb

Looking up at the star filled sky on a dark night can fill us with wonder and awe. But what do we do, if we don't know the names of the stars or how they move? This book is a guide which can help us get to know the stars and understand their movements without losing our sense of wonder.

The Sun, Moon and stars move across the sky. But we are told this is not true, that the Earth rotates and makes it seem like that. What is true?

It depends whether we think of the Earth at the centre, with *everything* moving around it (geocentric), or with the Sun at the centre (heliocentric). Both viewpoints have merit. One relates to our immediate experience of how we see the stars and celestial bodies moving; the other relates to our understanding, giving us a clear overview. The basics of the heliocentric system are part of our general knowledge today, but far too little is done to help us apply and translate this understanding into what we can observe ourselves. This book tries to bridge the gap between our first-hand observations and our theoretical knowledge.

The descriptions in this book are primarily of what can be seen with the unaided eye. Only occasionally is something described that needs binoculars or a telescope to be seen. Therefore the planets Uranus and Neptune (and the dwarf planet of Pluto) are only mentioned in passing. We do not go into astrophysics, and numbers are only given where they help form a picture.

During the author's time as a teacher he found that astronomy was often an unloved poor cousin to other subjects. This book will also help teachers to better understand astronomy, and be able to bring it to life for their students.

Introduction

This book begins with the phenomena of the heavens as we see them from the earth, that is, geocentrically. This is followed by an explanation using the Copernican or heliocentric system, as if we were viewing the Earth from a point in space. At the end of each section is a summary. Readers with some familiarity of the subject can just read the summaries until they find unfamiliar material.

There are four parts. First, the movement of the stars, primarily described as seen from a latitude of 50° North — that is, the south coast of England or the southern cities of Canada — but also describing their movement for other latitudes, including the southern hemisphere. In the other parts the movement of the Sun, the Moon and planets (as well as comets and meteors) are covered.

Areas of latitude are sometimes grouped as follows:
- *Polar latitudes:* north or south of about 65°
- *Temperate latitudes:* between about 25° and 65° (north or south)
- *Tropical latitudes* or *tropics:* between about 25°N and 25°S.
- *Higher latitudes* simply mean further north (or in the southern hemisphere, further south).

The illustrations showing south in the centre, and west and east to left and right, should really be curved into a semicircle so that west and east are opposite, and the top should be curved into a dome. Of course this isn't possible in a book, and that leads to distortions. Nevertheless the images can be a help to understand the movements or find constellations.

Illustrations showing the Earth and Sun from a distant viewpoint are not to scale. The Earth is often shown much larger in relation to the Sun than in reality. (The real relation of size and distance of Sun and Earth is similar to a football in the centre of the soccer pitch and a pea about halfway to the goal.) To illustrate other points, some features are enlarged or exaggerated, for instance of the Milky Way is drawn brighter than in reality.

Numbers are given as rounded figures, sufficiently precise to aid understanding. Of course distances in the sky cannot be given in miles or kilometres; they are given in degrees. Times are local times, ignoring daylight saving time.

Binoculars can be helpful. Their power is usually given as two figures, for instance 8 x 40. The first figure gives the magnification (in this

Measures in the sky

Following are true for most people with outstretched arm:
- 1° width of little finger
- 5° width of middle three fingers close together
- 10° width of hand at knuckles
- 15° distance between outstretched second and little finger
- 25° distance between outstretched thumb and little finger

case eightfold), the second figure is the diameter of the lens in millimetres. To avoid the view being shaky with hand-held binoculars, the magnification should not be higher than 8. For night-time viewing as large a lens as possible is best (40 is adequate, but 50 gives a brighter image). For viewing overhead objects, it is helpful to lie on a recliner.

PART I

THE STARS

From where can we best see the stars?

To see the stars well, we not only need to go out during a cloudless night, but need to find somewhere dark with a clear all-round view. This is obviously difficult in a city where the nearby houses block the view and street lighting makes our eyes insensitive to the fainter light of the stars. Badly directed lighting also illuminates the haze that usually hangs over a city. Astronomers call this light pollution. There is growing awareness and efforts to reduce it, and an international dark sky movement is campaigning to make the stars more visible at night.

Things are a bit better around the edges of cities, or in a large, unlit park. Stargazers should search out places where there are as few distractions from lights as possible. Try to turn your back to the illuminated city haze, and at least half the sky will be visible. It is also easier to see the stars on a moonless night than with the glare of a full Moon.

Out in the countryside, far from lights, is good, and if you find such a place not too far from your home you will be able to visit it regularly. The ideal situation is high in the mountains or in the dry air of the desert.

Once your eyes have got used to the dark, and it is a clear, moonless night, about 2500 stars can be seen. Counting a similar number of stars below the horizon, this means that in total about 5000 stars can be seen with the unaided eye. (Some estimates put this figure a little higher.)

At first the view can be confusing without any firm reference points. But just as we can find our way around our neighbourhood using familiar landmarks, so we can orientate ourselves in the sky. Groups of stars and some individual stars have names that we can get to know. However, unlike our local landmarks, the stars in the sky move — many rising and setting like the Sun or Moon. In summer we see different stars than in winter. However, in relation to one another they keep their positions, and therefore are also called 'fixed' stars. One star, though, always keeps its position, and that is the Pole Star (or Polaris). If we are in the northern hemisphere, we can begin by finding that star and thus finding where north is.

The Pole Star (Polaris)

The best known constellation in the northern hemisphere is Ursa Major, the Great Bear, the Plough or Big Dipper, with its seven equally bright stars (Figure 1). It can always be found in temperate northern latitudes where it never sets. Four stars can be seen as outlining the pan of the Dipper, and the three other stars are like the bent handle of the Dipper (or the shaft of the Plough).

The two stars opposite the handle point to the Pole Star at a distance of about five times as great as the distance between them. The Pole Star (or Polaris) always has the same position, standing still. Drop an imaginary line straight down from the Pole Star to the horizon, and that is north. Facing north, east is on our right, west on our left, and south is behind us. Finding the four directions, or getting our bearings, is the first step to recognising the stars in the sky.

The height of the Pole Star above the horizon, measured in degrees, is the latitude of our location. To accurately measure angles in the sky sextant is used (or a Jacob's staff before the eighteenth century). In London (and Calgary, Alberta) the Pole Star is about 51° high, while in Quebec City (or Budapest, Hungary) it is about 47°; in Philadelphia, Pennsylvania (or Madrid, Spain, or Beijing) it is about 40°; and in New Orleans (or Cairo, Egypt, or Lhasa, Tibet) it is only about 30° high.

Figure 1. The Great Bear or Big Dipper in the evening in November. Extend the line of the last two stars for about five times the distance between them to find the Pole Star, which is above the north point of the horizon. The height (*h*) of the Pole Star above the horizon is the same as the latitude of the place of observation.

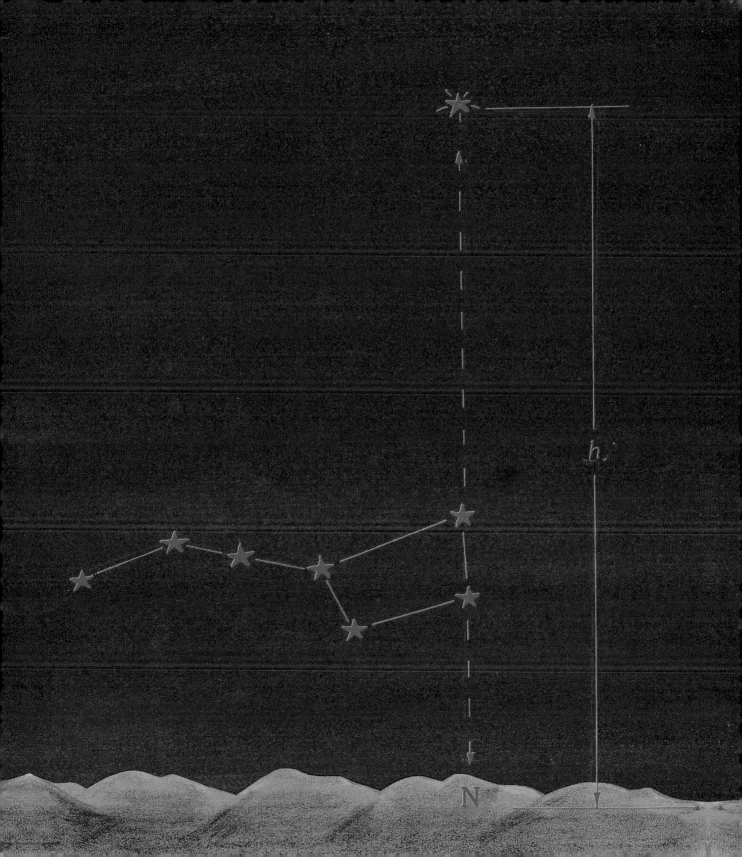

N

h

1. The Daily Movement

The northern hemisphere

We shall begin our description of the daily movement of the stars as seen in northern latitudes of about 40° to 50°.

Looking north

The Great Bear or Big Dipper is not always low down, as shown in Figure 1. It is a good exercise on a clear night to go out every hour looking north, and to see how the Dipper moves around the Pole Star in an anticlockwise direction. After, say, 6 hours, we can see that it would take 24 hours to move right around (Figure 2). The Romans called this constellation the seven threshing oxen — the cattle moved around a central pole pulling the threshing boards over the cereals to separate the grain from the straw and chaff.

There are other stars and constellations which never set. They circle around the Pole Star and are called *circumpolar stars*. The higher the Pole Star is — and thus the further north we are — the bigger the circle of stars that never set.

A simple aid to help understanding this movement is to draw the seven stars of the Great Bear on the inside of an open umbrella. Point the top of the umbrella towards the Pole Star, and turn it anticlockwise to see how the stars move around the Pole Star.

Looking south

Turning around and leaving the Pole Star behind us, we are facing south. The daily movement of the stars can be followed here as well. Facing south the stars move clockwise. They rise on our left in an easterly direction and set on our right in a westerly direction. Between these two they reach their highest point in the south; this is called their *culmination*. Their path is the shape of an arc. It is the same movement as we saw facing north, but because we are looking south, it appears to be in the opposite direction.

If the points of culmination are all joined they form a line at a right angle to the horizon, going straight up to the *zenith* (the point directly above our heads) and on to the Pole Star and the north point on the horizon behind

Figure 2. The course of Ursa Major, the Great Bear, over 24 hours is anticlockwise.

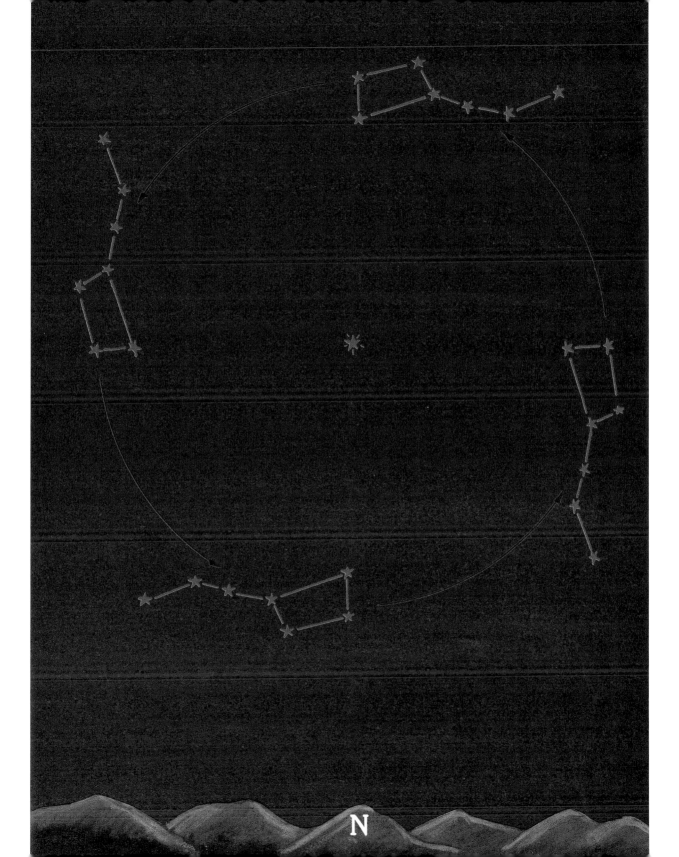

N

us. Astronomers call this invisible semicircle the *meridian*. It continues below the horizon through the *nadir* (the point exactly below us, opposite the zenith) and back to the south point, forming a complete circle. The meridian is exactly perpendicular to the horizon and is a *great circle* (where we are in the centre, so that it appears absolutely straight wherever we look). All the stars in the course of their daily movement culminate as they pass through the meridian.

As an example let's look at the constellation of Orion, the hunter, that is visible in winter in the northern hemisphere. It is a prominent constellation with the three close stars of the belt, some distance above the two bright stars of the shoulders and the less bright star of the head; about the same distance below are two bright stars of the feet (Figure 3).

Great circles

We can distinguish what in spherical geometry is called a *great circle* from other circles: the horizon is a great circle. We stand in its centre, and wherever we look it is straight and flat. The circles the stars of the Dipper make around the Pole Star are all smaller circles.

The arcs of Orion's stars are centred on the *celestial south pole*. This is a point below the horizon as far as the Pole Star is above it, and exactly opposite it.

The most northerly star of the three in Orion's belt rises exactly in the east and sets exactly in the west twelve hours later. This is because that star lies precisely on the *celestial equator*. It divides the northern celestial hemisphere from the southern and, like the Earth's equator, lies between the two poles. Like the meridian, the celestial equator is a great circle, and at any time half of it is above the horizon. It goes exactly through the east and west points of the horizon, and is tilted from the zenith by the same number of degrees as our latitude.

If we face the north we can only see the stars of the northern hemisphere, but if we look south, the stars above the celestial equator are part of the northern stars and those below belong to the southern hemisphere. The star in the belt of Orion, and all the other stars on the equator are the opposite of the stationary Pole Star, for in their daily movement they follow the longest path (great circle) and move the fastest.

Only the stars on the equator rise exactly in the east and set exactly in the west twelve hours later. The stars above the celestial equator, those of the northern hemisphere, rise to the north of east and set to the north of west. They can be seen for longer than twelve hours. The further a star is from the celestial equator, the longer it will be seen. The circumpolar stars are far enough from the celestial equator to be seen above the horizon all the time. The stars

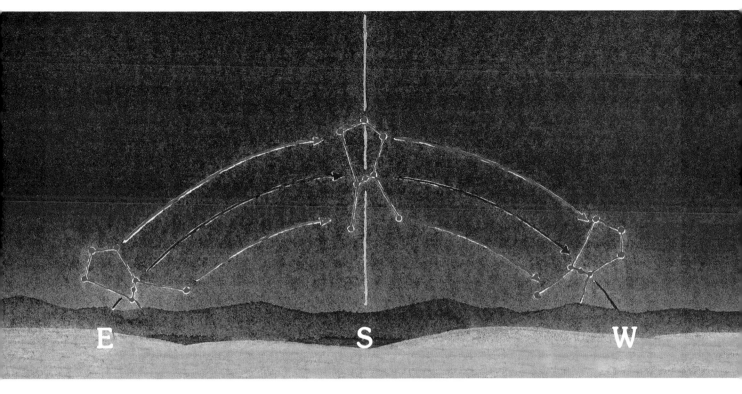

Figure 3. Orion rising, culminating and setting.

below the equator belong to the southern celestial hemisphere, and rise and set to the south of east and west. They are above the horizon for less than twelve hours. The further from the equator, the shorter they appear above the horizon, until the circle of stars which just appear on the horizon for a moment; below that are the southern circumpolar stars which are below the horizon all the time. (In Figure 3 the paths of the northern stars appear to be greater than the great circle of the celestial equator. This is an inevitable distortion when trying to depict part of a sphere onto a flat image.)

When rising in the east, Orion is tilted to the left; at culmination in the south the hunter appears upright; and then tilts to the right on setting (Figure 3). During the course of a night all constellations turn like we have seen with the Great Bear and Orion. To find our way around the sky at night it is good to remember how this turning movement works.

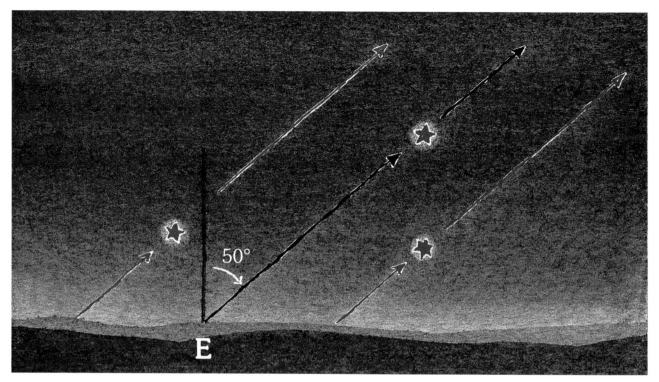

Figure 4. Slanted rising of stars in the east, shown for latitude 50°N.

Looking east and west

If we look east we see a part of the horizon where stars are continually rising. We can see that they do not simply rise vertically, but at an angle. To see this clearly, note a point on the horizon where you can see a star. Look a little later and you can see the star has moved to the right and gained height (Figure 4). In the west it is more difficult to see this because stars are continually setting. We need to find a higher object on the horizon like a tree, steeple or mast and then observe some minutes later how the star has become lower and moved to the right (Figure 5).

The angle at which the stars rise and set depends on our latitude. The angle from the vertical is the same as the latitude. So the further north we are the greater this angle is, and thus the smaller the angle to the horizontal, or the shallower the rising and setting. The closer we are to the equator the more steep the rising and setting becomes.

Higher northern latitudes

As we travel north, the Pole Star appears higher. Every 111 km (70 miles) it rises by 1°. This figure can be quickly calculated from the size of the earth, for 1 km was originally defined as 1/10 000 of the distance from the North Pole to the equator (more recent and accurate

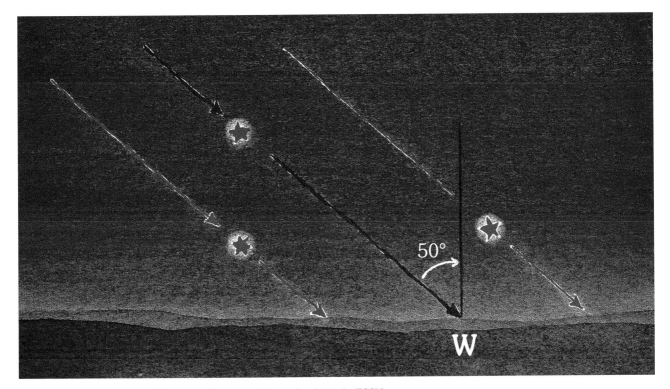

Figure 5. Slanted setting of stars in the west, shown for latitude 50°N.

measurements have shown the distance to be slightly more than 10 000 km). As there are 90° we can divide 10 000 by 90 to get 111.

In Britain, Southampton is at a latitude of 51°N and Orkney is at 59°N — a difference of 8°, equivalent to a direct distance of about 888 km (550 miles). In North America, Washington DC or Sacramento CA are at 39°N, and Quebec City or Seattle WA are at 47° — a similar north-south difference. So the Pole Star will be 8° higher in these northern cities than in the southern ones.

As we go north the number of stars that become circumpolar increases. The angle of the celestial equator to the horizon becomes more shallow, so the rising and setting lines are less steep.

The North Pole

If we continue north we cross the ice of the Arctic Ocean. Eventually we would reach the North Pole where, corresponding to latitude 90°, the Pole Star is at the zenith, 90° above the horizon, and the Great Bear circles overhead. The celestial equator exactly matches the horizon. East and west disappear, for all directions are south. All the stars are circumpolar, with each star circling horizontally from left to right at an unvarying height above the horizon. Only the stars of the northern

Figure 6. At the North Pole only the northern part of the zodiac can be seen (see Figure 24 for zodiac symbols). All stars circle horizontally from left to right.

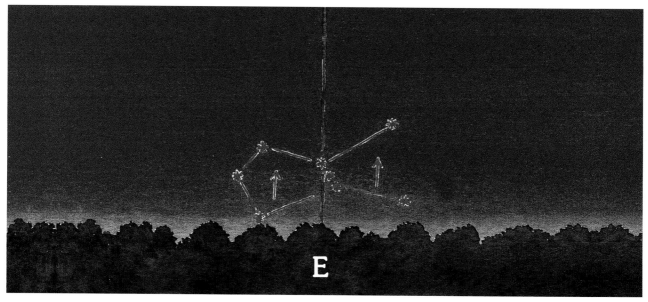

Figure 7. Orion rising at the equator.

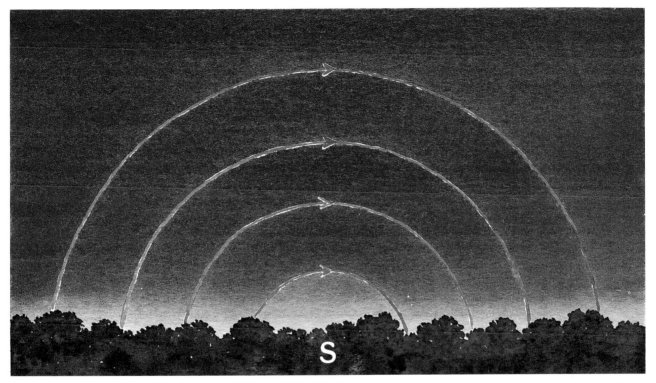

Figure 8. Paths of the stars looking south from the equator. Note that all the arcs are semicircles: half above and half below the horizon.

celestial hemisphere can be seen, so only the northern half of Orion is visible (Figure 6).

The equator

As we travel south the Pole Star is lower on the horizon. Fewer stars are circumpolar, and more stars appear in the south. In Philadelphia, Pennsylvania (40°N) more than half the stars of the southern hemisphere are visible. The angle of rising and setting stars is steeper.

If we continue southwards as far as the equator, the Pole Star will be on the horizon.

There are no circumpolar stars, and all stars rise and set vertically (Figure 7). It is possible to see all stars of the northern and southern celestial hemispheres during the course of a night.

The celestial equator goes through the east point of the horizon, the zenith overhead, the west point, and below us through the nadir. Looking south the stars move clockwise, each star on a semicircular path, so that each star is above the horizon for 12 hours and below the horizon for 12 hours (Figure 8). Looking north the stars move anticlockwise.

The southern hemisphere

In populated parts of the southern hemisphere all the constellations discussed above, except for the circumpolar ones, are also visible. However, they appear the other way up. Visitors familiar with the stars at home find they are upside down in the opposite hemisphere. In addition, looking south we find new circumpolar stars circling around the south celestial pole clockwise — in the opposite direction of the circumpolar stars around the north pole.

In the northern hemisphere we can see Orion in winter, and we might think that to see it in summer we could go to the southern hemisphere where it is winter. But that is not so! In June and July, the southern winter, the Sun is in the constellation of Taurus and Gemini, close to Orion, and so we cannot see Orion at that time of year, no matter where we go. In the southern hemisphere Orion is a summer constellation, visible in December.

Antarctica and the South Pole

In Antarctica we can see a great number of circumpolar stars, and only a few of the northern hemisphere constellations just above the northern horizon. At the South Pole we have a similar situation to the North Pole: we can only see half the stars, but all the stars that can never be seen from the North Pole.

Photographing the movement of the stars

To photograph the stars, the camera needs to be mounted on a tripod. If the shutter is opened for about a minute (*time exposure*, best done with a wire release) the stars appear as points. A normal camera (50 mm lens) will produce good photographs if the sky is dark enough. Experiment with different exposure times and keep notes.

With an exposure time of 5 or 10 minutes small lines appear; the movement of the stars becomes noticeable. To get really clear lines of movement an exposure time of at least one hour is needed. If you point the camera towards the Pole Star, you can see the seven bright lines of the Great Bear's stars. Other stars will make greater or smaller circles. Only the Pole Star hardly moves. Stars close to the Pole Star make small circles (Figure 9). This photograph shows how the stars move in circles. Even the Pole Star makes a tiny circle, as it is, in fact, not exactly in the celestial north pole.

The whole thing looks quite different if we point our camera at Orion. The stars appear to move in straight lines (Figure 10). We can make out the stars of Orion's belt that are on the celestial equator. But in Figure 3 the celestial equator is shown as an arc, not a straight line. Both images are correct. In the photograph, the camera is at the centre of the great circle of the celestial equator, and only shows a narrow view to the south. The drawn picture is distorted (as mentioned earlier) as the east and west points are actually opposite each other to our left and right.

Figure 9. Time exposure of the stars around the Pole Star. [Martin Rietze]

Figure 10. Time exposure of the constellation of Orion. The three stars of the belt can be seen clearly in the upper part. [Martin Rietze]

If we take photographs east or west we would see the slanting rising or setting lines. Try different directions yourself.

In the photographs we can see that the lines are different colours. The stars really are different colours, which we can only just discern with the naked eye, but it is shown more clearly in the photographs.

The Daily Movement: Summary

- The daily movement affects all stars, and we see their rising in the east and setting in the west.
- In northern temperate latitudes the stars rise in an easterly direction, slanting towards the south, culminate in the south and then set diagonally down towards west. In the southern hemisphere they rise slanting to the north, culminate in the north and then set towards the west.
- After about 24 hours they are back in the same position. Looking south their movement is clockwise, looking north the movement is anticlockwise (this is true wherever we are on the Earth).
- The stars circle the celestial pole (in the northern hemisphere the Pole Star) whose angle above the horizon is the same as our latitude. The stars closest to the pole never set and are called *circumpolar stars*.

(Opposite these are circumpolar stars that we never see because they are always *below* the horizon.)

- The *celestial equator* rises exactly in the east, culminates in the south (in the north in the southern hemisphere) and sets exactly in the west. It divides the sky into a northern and a southern celestial hemisphere. The northernmost star of Orion's belt is fairly precisely on the celestial equator.
- The *meridian* goes from the south point of the horizon through the zenith (the point exactly overhead) to the north point of the horizon, and then below the horizon through the nadir (the point exactly below us) back to the south point. The stars culminate when passing the meridian.

2. The Rotation of the Earth

The *celestial sphere* — that is, the entire starry sky as we see it from where we stand — rotates around an invisible axis that passes through the Pole Star and the south pole. The stars appear to be fixed to this celestial sphere, hence they are also called 'fixed stars' (in contrast to the 'wandering stars', the planets).

Until now we have described mainly the daily motion of the stars. What causes them to appear this way? Do they really all turn around the Earth without losing their places?

In very early times the Earth was thought to be flat, with the stars and planets moving on spheres around it. In ancient Greece astronomers knew the Earth was round (and even had a reasonably accurate idea of its size), but most of them still believed the Earth to be standing still. Through the Renaissance the idea that the Earth turned around its axis in 24 hours and the stars, Sun and Moon remained (relatively) still became increasingly accepted.

Regardless of whether the Earth stands still and the stars move around it, or the stars are still and the Earth rotates, the movements appear to us as we have described them. We have probably all been on a train, looking at another train alongside, and thought we were moving, only to realise (when we looked at the platform) that it was the other train that moved in the opposite direction. When we look at the stars, however, we don't have a convenient platform to check our observation!

If we imagine the Earth is moving, we imagine it from a point outside the Earth — something we can only do in our thoughts. It was Nicolaus Copernicus (1473–1543) who first brought this idea to the wider world. His Copernican system (as we now call it) saw the Sun in the centre, with the planets and Earth circling around it.

The Earth is a rotating sphere. Conventionally, in models and maps, north is shown at the top (of course, we have to remember that wherever people live on the Earth, they feel the ground below them and the sky above them). To determine which way the Earth rotates (or to check if the arrows in Figure 11 are correct), the stars have to set in the west; that means an observer on the Earth has to be moved away from them so they disappear below the horizon. As Figure 11 shows, the Earth and the observer move anticlockwise (seen from the north) for this to happen.

Another phenomena can also be explained with this theory. As the Pole Star is in line

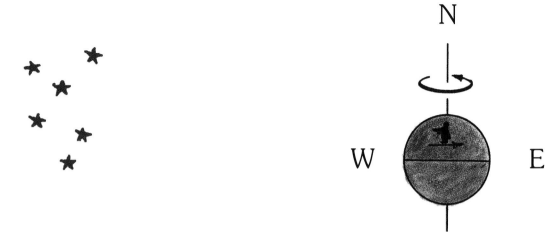

Figure 11. Because of the Earth's anticlockwise rotation, the stars appear to set in the west.

with the Earth's axis, it appears to stand still as the Earth rotates. It is so far away from the Earth that the different position of an observer twelve hours later (points *A* and *B* in Figure 12) cannot be measured. We know the angle of the Pole Star above the horizon is the same as the latitude. This too can be seen in Figure 12. The (blue) angle of the Pole Star to horizon is equivalent to the (green) angle of latitude (as the arms of the angles remain perpendicular).

We can see that circumpolar stars are those close to the pole that are within the extended horizon (blue) line of the observer, this always above the horizon.

The slant of the rising and setting of stars, Sun and Moon can also be explained by the Copernican theory. Someone standing at the Earth's equator is perpendicular (at right angles) to the Earth's axis, and thus the stars appear to rise and set vertically to the horizon (Figure 13).

For someone in temperate latitudes, the horizon is at an angle to the axis of the Earth and so the stars appear to rise and set at an angle to the horizon.

The great pendulum

Can the Copernican theory be proven? There are two methods. The second, the parallax of the stars will be discussed on page 78.

The first method, which is simpler and easier to see, is one that the French physicist Léon Foucault demonstrated in 1851. He suspended a pendulum from the Pantheon in Paris. The weight was attached to a wire 67 metres (220 ft) long, and the position where it began to swing was very carefully marked. After some hours the direction of the swinging changed. The only plausible explanation is that the plane

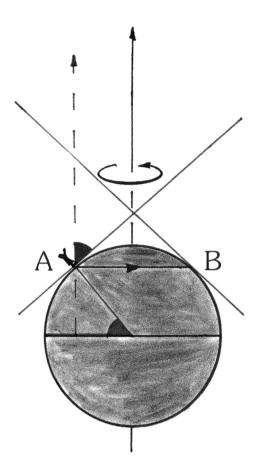

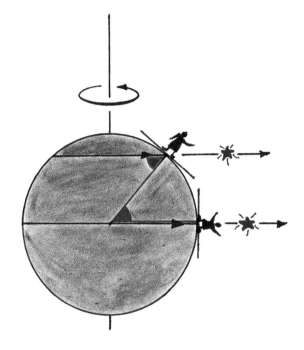

Figure 12. The blue angle shows the angle of the Pole Star above the blue horizon. The green angle is the latitude of the observer. These two angles will always be equal, no matter where the observer is. (Because of the great distance of the Pole Star from the Earth it will appear parallel to the Earth's axis.) All stars between the blue horizons are circumpolar.

Figure 13. The two rising stars must be imagined infinitely far away. For the observer on the equator they appear to rise vertically, but for someone in temperate latitudes they appear to rise at an angle to the horizon.

of the pendulum's swing remained constant, and the Earth (and the corresponding marks on the floor) rotated beneath it, making it appear as if the pendulum was turning. This was the famous experiment of Foucault's pendulum.

At the equator this experiment would not show any rotation, while at the Poles the pendulum would make a complete rotation in 24 hours, moving through 15° every hour. Between the equator and the Poles the hourly displacement is less than 15° and thus the complete rotation takes longer. Such a pendulum can be found in many museums or observatories around the world.

There are other phenomena of nature connected to the rotation of the Earth. A meteorological chart will show the winds around an anticyclone (a high pressure area) circling clockwise, and around a cyclone (a low pressure area) circling anticlockwise. But this is only true of the northern hemisphere — in the southern hemisphere it is the other way round. This is caused by the rotation of the Earth.

The Rotation of the Earth: Summary

- It is easy for us to imagine the Earth rotating and the stars standing still, following Copernicus' idea. Everything we have observed so far can be explained in this way.
- However in our *everyday* experience we speak of sunrise and sunset, and this corresponds to what we *see*. Astronomers speak of the 'apparent' motion of the stars. This does not imply, as we might think, that it is not *real* — it is simply using the word in its literal sense of 'as it appears'.

3. Constellations

The stars we can see change not only during the course of a night, but also during the course of a year. The summer constellations are not the same as the ones we see in winter, and in spring and autumn different stars are visible. So as well as the daily motion, there is an annual one. This is caused by the fact that *every week* each star rises and sets about half an hour earlier.

Northern hemisphere

In springtime we see the Spring Triangle of (Figure 15); throughout summer we can see the Summer Triangle (Figure 16); in autumn there is the square of Pegasus (Figure 17); and in winter the Winter Hexagon. These four groups will help us find all other visible constellations throughout the year. To the north we can always see the same constellations — the circumpolar stars — though their orientation changes.

It is not easy to find our way around the stars to begin with. Following descriptions will be a help, and more detailed books and star charts, like the *Stargazers' Almanac,* are available. A planisphere (rotating star chart) can also help, though they are often very small.

Winter constellations
Looking south around midnight at Christmas we see six bright stars of six different constellations forming a slightly irregular hexagon (Figure 14). They are Rigel in Orion, Sirius in Canis Major (the Great Dog), Procyon in Canis Minor (the Little Dog), Pollux in Gemini (the Twins), Capella in Auriga (the Charioteer) and Aldebaran in Taurus (the Bull).

Beginning with Orion, the Hunter, which we've already met, we see the two feet, the two broad shoulder stars, the slightly dimmer head between them, and the three stars of his belt. From his belt hangs a sword consisting of a number of stars and a nebula that can be seen with binoculars. To his right are a number of fainter stars that form a bow at shoulder height. Some pictures show Orion's left arm raised, brandishing a club. (There are no fixed images for the constellations, so we can let our imagination complete the picture.) Many of the names of stars we use today come from Arabic, and Rigel, his right foot (from our viewpoint), means 'foot' in Arabic. His left shoulder is

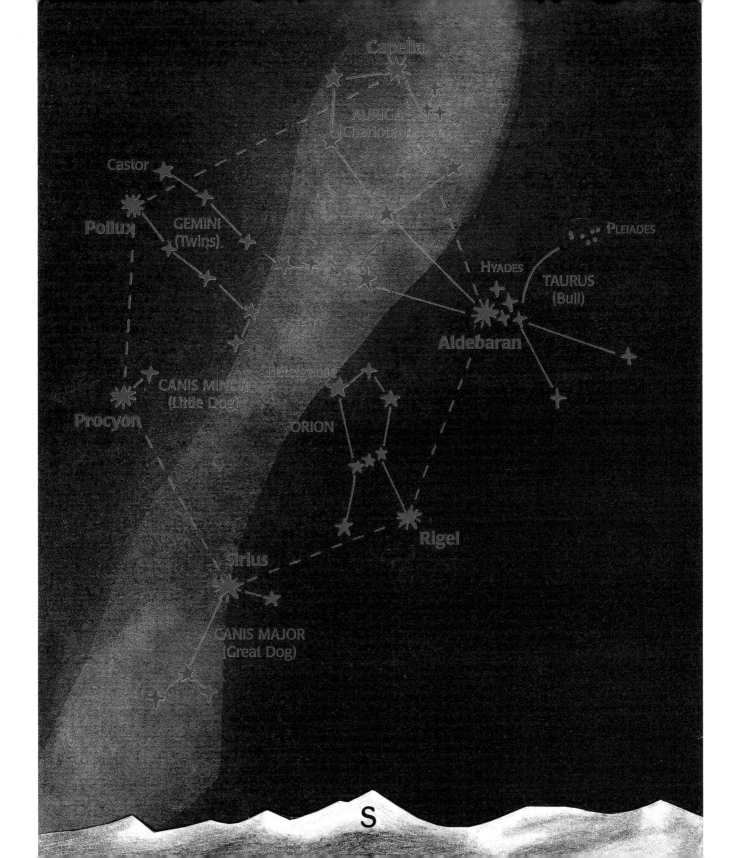

Betelgeuse (a corruption of Arabic meaning 'hand').

A hunter often has dogs, and we can see this in the sky. The next star of the hexagon is Sirius in Canis Major, the Great Dog. Sirius is the brightest star in the sky, glistening with all colours into a brilliant white. The constellation is formed from the stars below and a bright one to the left. Sirius is the head, and so the dog appears to be leaping up. At Epiphany (January 6) Sirius is exactly in the south at midnight. It is the lowest star of the hexagon.

The third star is Procyon, the brightest of Canis Minor, the Little Dog. It can be found about one hand span above and a little to the left of Sirius. The only other significant star in that constellation is to the right and a little higher. The name Procyon comes from the Greek meaning 'preceding the dog' as it always rises before Sirius in the Great Dog.

Higher in the sky are two stars, Castor and Pollux, the heads of Gemini, the Twins. The lower star, Pollux, is slightly brighter and is one of the hexagon. The bodies of the Twins can be seen as two lines of four stars each to the right and downwards. There are also stars at their feet. With a very clear sky some fainter stars indicating the arms can be seen. This constellation is part of the zodiac which we shall look at later.

The highest star of the hexagon, Capella, is almost overhead, and is part of the constellation of Auriga, the Charioteer. This looks a bit like an elongated five-sided shape with Capella at the top. To the right of Capella are three fainter stars that are sometimes seen as the goat-kids. North of latitude 44°, Capella is circumpolar.

The sixth star of the hexagon is below and to the right of Capella. Aldebaran is a reddish star and is the eye of Taurus, the Bull. It is at the top of a V-shape of five stars, the Hyades, which together form the head of the Bull. The two long horns of the Bull are in the direction of Auriga; one of the horns is common to that constellation, though modern astronomers have assigned it exclusively to Taurus. To the right above the Hyades is the cluster of the Pleiades or Seven Sisters. Despite their name usually only six stars can be discerned. Their arrangement, similar to the Great Bear, can be seen more clearly with binoculars. The Pleiades are the back of the Bull, and there are some stars below which form the front legs. Taurus was seen as Zeus in the guise of a bull who carried off Princess Europa to Crete.

As was said earlier, these constellations culminate, appearing like this, at Christmas around midnight. They rise around dusk and set around dawn, dominating the sky throughout the night. By the end of January they rise earlier, culminate at 10 pm, and set well before dawn. By the end of February they culminate as early as 8 pm, setting in the second part of the night. By the end of March, when they culminate at 6 pm, we see them setting after nightfall. In summer they cross the sky during the day when the light of the Sun makes them invisible. By the end of September they rise before dawn, culminating at 6 am when it is light. Over the following months they rise and then culminate 2 hours earlier each month.

Figure 14. The Winter Hexagon in the south, at Christmas around midnight.

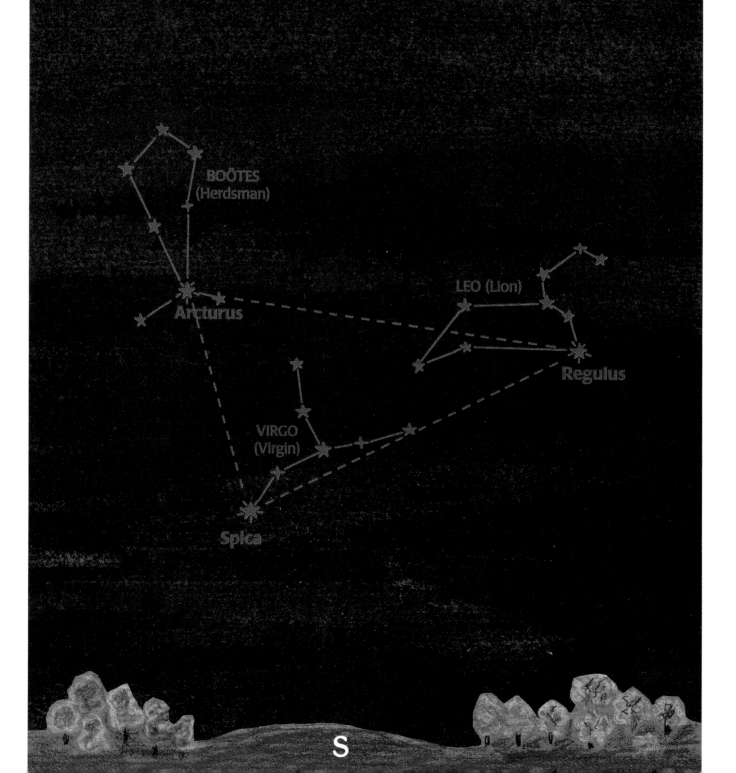

To the left of Gemini is the small, faint constellation of Cancer, the Crab. It is a triangle with the star cluster Praesepe, the Manger, in its centre (best viewed with binoculars). Further to the left the magnificent constellation of Leo, the Lion, rises. To the right of Taurus are three stars that form Aries, the Ram. The two brighter stars are its head and horn. This is the ram of the Golden Fleece. Further to the right the stars of Pisces, the Fishes, are setting. Pisces consists of a number of less bright stars.

Below Orion is Lepus, the Hare, a small constellation of fairly bright stars that is outshone by the brighter stars of Orion. Below Aries is Cetus, the Whale, with Mira, a star of variable brightness. Cetus was the sea monster who was about to devour Andromeda before she was rescued by Perseus. Perseus can be seen above Taurus, to the right of Auriga. Perseus carries the hideous head of Medusa, a Gorgon or female monster. The head is represented by Algol, another variable star. Every two-and-a-half days, Algol becomes noticeably dimmer for a few hours.

Also above Aries is the little constellation of Triangulum, the Triangle. This elongated triangle traditionally pointed to the beginning of the zodiac.

Spring constellations

The Spring Triangle consists of the three stars Regulus in Leo, Arcturus in Boötes, and Spica in Virgo (Figure 15). Leo and Virgo are part of the zodiac. This triangle culminates at midnight

Figure 15. The Spring Triangle in the south at midnight in April.

in April. In February it rises at midnight, but in December is visible before dawn. It can be seen in the evening until July when it sets at dusk.

Regulus, meaning 'little king', is the brightest star in Leo, the Lion. It is at the lower right of an elongated four-sided figure. Above Regulus is an arc of stars showing the great head of the lion, and to the left are the hindquarters and tail. The Lion is one of the constellations where it is easy to recognise the creature from the grouping of the stars. According to legend it is the Lion of Nemes that Heracles had to conquer as his first task. Above Leo is the Great Bear.

Arcturus is the main star of Boötes (pronounced *boh-ohtiz,* with the emphasis on the second 'oh'). This constellation can also be found by following the curve of the three stars of the 'handle' of the Big Dipper, the tail of the Great Bear. Boötes is also called the Herdsman, as he drives the Great Bear in front of him.

Following the sweep of the tail of the Great Bear (or handle of the Dipper) through Arcturus, we find Spica, the main star of Virgo, the Virgin. Spica is the ear of wheat which the Virgin carries. Virgo is a very long constellation consisting of many fainter stars, and it is not easy to recognise a human figure. You can see a lying down Y-shape with Spica at the foot of the Y. Virgo is part of the zodiac, and between Virgo and Leo lies the autumn equinox which we shall mention later (page 69).

To the west, Auriga and Gemini can be seen setting, reminding us of winter. To the east, left of Virgo, the four dim stars of Libra can be seen. It is easy to recognise an old fashioned pair of scales with the two pans hanging from a beam. Below Virgo is Corvus, the Crow, and Crater, the Cup. Hydra, the Water Snake, consists of a long winding string of faint stars. Around midnight Ophiuchus, the Serpent Bearer, and Serpens, the Serpent, rise. These constellations are intertwined. Above them is Hercules, and between Hercules and Boötes is Corona (Borealis), the Crown, a beautiful small, semi-circular constellation; its brightest star is Gemma, the jewel. (There is also a Southern Crown, Corona Australis.) Some time later the stars of the Summer Triangle rise.

Summer constellations

The three stars Vega in Lyra, Deneb in Cygnus and Atair in Aquila form the Summer Triangle (Figure 16). In May the triangle rises around midnight, and at the end of June around 10 pm. In July it culminates at midnight. In the following months, as sunset becomes earlier, it can still be seen in the early evening until January.

The bluish-white Vega, at the right hand point of the triangle, has some much fainter stars nearby, forming Lyra, the Lyre. On a clear night they can be seen to form a beautiful symmetrical U-form with down-turned ends. Vega is the bright star at the right hand top of the U. The constellation represents Apollo's lyre with which Orpheus went to the underworld.

Cygnus, the Swan, is by contrast a very big constellation. Its brightest star, Deneb, is the tail. Four stars, spaced evenly almost in a straight line, form the body and long neck. Two bent back wings can be seen on either side of the body. The great bird is flying with outstretched wings and neck downwards. These stars of the Swan are sometimes also called the Northern Cross.

Atair is the eye of Aquila, the Eagle, which is flying in the opposite direction of the Swan. Two stars are close by and it is from these that the wings stretch. They are not as symmetrical as those of the Swan: it is almost as if the right wing is pulled in a little to avoid colliding with the Swan. The body can be clearly seen — the Eagle has a short neck and a comparatively long tail.

Within the Summer Triangle is the tiny constellation of Sagitta, the Arrow. Above Aquila and to the left is Delphinus, the Dolphin, another small constellation. The Summer Triangle is almost completely within the Milky Way.

During summer nights low in the south, Scorpius and Sagittarius can be seen; these two constellations are part of the zodiac. Scorpius' main star is the reddish Antares. The pincers are to the right, almost enclosing Libra, and to the left and below is the curved tail. This beautiful constellation can only be seen properly south of latitude 45° when the tail is above the horizon.

To the left of Scorpius is Sagittarius, the Archer, a compressed constellation with a bow and arrow pointing at Scorpius. Sagittarius is a centaur, a wild creature, half man and half horse. Seen from northern latitudes this part of the zodiac is very low, but it has many bright stars.

Between Scorpius and Sagittarius is the winter solstice point. Just below them is the Southern Crown, Corona Australis; you need to be south of latitude 40°N to see it clearly.

Figure 16. The Summer Triangle in the southeast at midnight around the end of June.

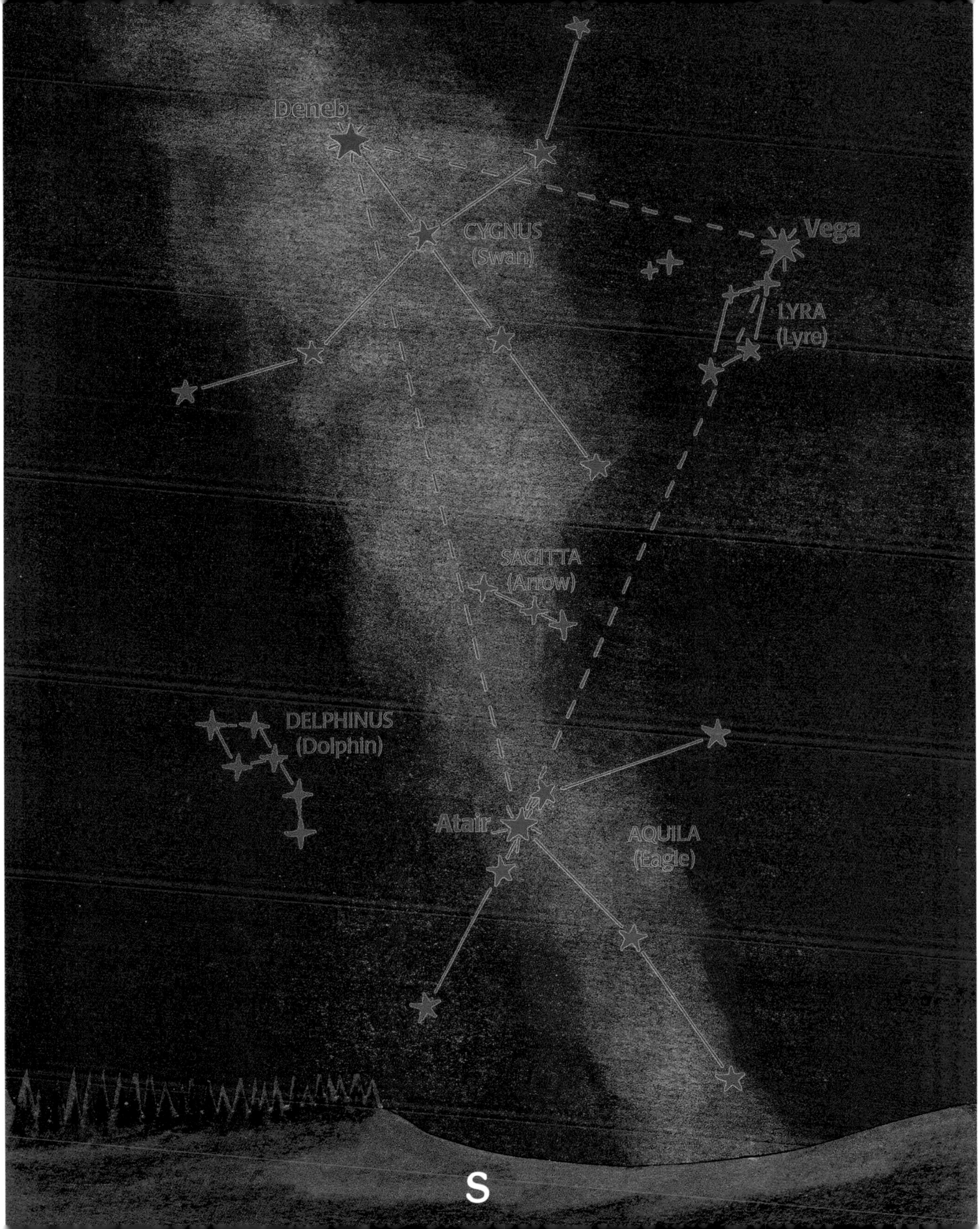

Deneb

CYGNUS
(Swan)

Vega

LYRA
(Lyre)

SAGITTA
(Arrow)

DELPHINUS
(Dolphin)

Atair

AQUILA
(Eagle)

S

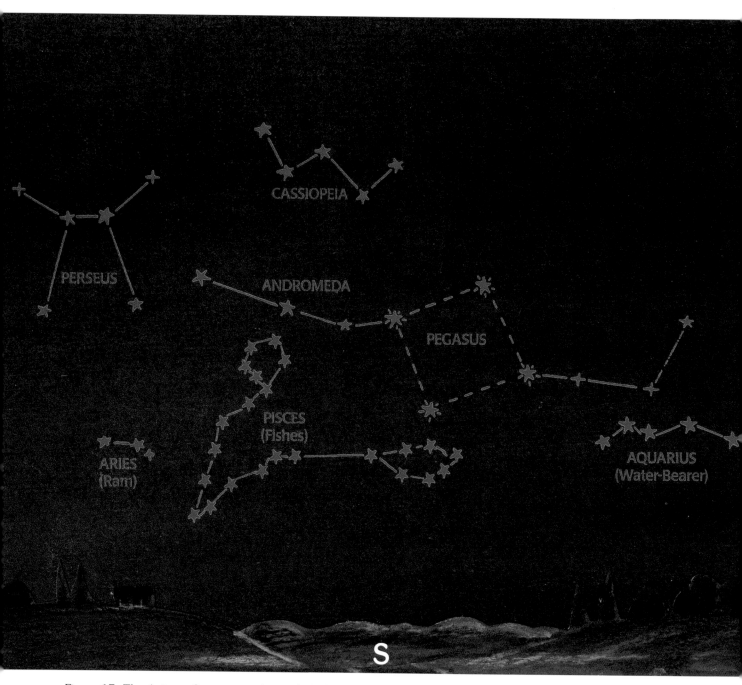

Figure 17. The Autumn Square, together with Andromeda, Cassiopeia, and Perseus, culminating in the south at midnight in late September.

Above Scorpius and Sagittarius, Ophiuchus and Hydra can be clearly seen, and higher up is Hercules.

Capricornus, the Sea Goat, is a more modest constellation, a kind of curved triangle with horns at the right.

Autumn constellations

The Autumn Square is not as impressive as the shapes of other seasons, but can be clearly seen (Figure 17). Its four stars are comparatively close together and all belong to one constellation Pegasus, the Winged Horse. It can be seen rising in the east in September evenings, culminating around midnight. It disappears in February in the west. The winged horse appears upside-down: the square is the body, the feet are above and to the right is the head.

The two upper stars of the square form a curved arc of stars to the left in ever increasing distances. The last star of the arc is part of Perseus. The stars in between are part of Andromeda. Above the centre star is the famous Andromeda Galaxy, the only galaxy visible with the naked eye in northern skies. It is worth having a closer look with binoculars. Andromeda is the princess who was chained to a rock, waiting to be sacrificed to a sea monster (Cetus, the Whale). Above her are Queen Cassiopeia, her mother, and King Cepheus, her father. Perseus comes to the rescue, turning the monster to stone with the head of Medusa.

Below Pegasus to the left and right are Pisces, the Fishes and Aquarius, the Water-Bearer. Aquarius is a large constellation but difficult to recognise. Look for the characteristic horizontal wavy line. To the left is the water jug, to the right his outstretched arm.

Pisces is even larger, consisting of many faint stars. On a clear night two ovals, the fish, can be seen connected by a ribbon. One fish points up, the other towards Aquarius.

Below Pisces is Cetus with Mira, and below Aquarius is Piscis Austrinus, the Southern Fish with the bright star Fomalhaut. Pisces and Aquarius are part of the zodiac and the spring equinox point lies between them. In the west, to their right, is Capricornus, the Sea Goat, and in the east Aries, the Ram.

Circumpolar stars

The further north we are on the Earth, the greater the number of circumpolar stars. The following description is for latitude 50° — that is, the south coast of England or the southern cities of Canada. Observers further south will see only constellations closer to the Pole Star. None of the zodiac constellations are circumpolar (except in uninhabited Arctic regions).

Ursa Major, the Great Bear, is low above the northern horizon during December evenings (Figures 1 and 2). During March evenings it is to the right of the Pole Star with the tail hanging down. In May and June it is high overhead, upside down. And in August and September it is to the left of the Pole Star, tail uppermost. During the course of a night the Great Bear repeats this movement, and so the annual movement in effect echoes the daily movement.

Figure 18 shows the Plough as part of Ursa Major, the Great Bear. The Plough or Big Dipper consists of the seven bright stars on the left. Above Mizar, the centre star of the handle

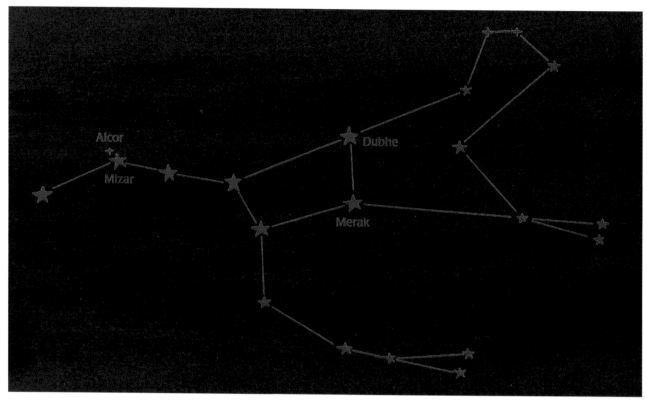

Figure 18. The Plough or Big Dipper is part of Ursa Major, the Great Bear. Dubhe and Merak are the two stars pointing to the Pole Star.

of the dipper or plough, is a little star called Alcor. These two stars are sometimes called the horse and rider. If you can see Alcor on a clear night, then you have good eyes (use binoculars if you can't). The Plough, as it is known in Britain, was sometimes also called the Wain (or Wagon), and in northern Europe it is still known as that. In North America it is known as the Big Dipper, and in Japan, China and southeast Asia it is known as the Northern Ladle. In ancient India these stars were known as Sapta Rishi, the seven holy sages.

The Plough or Big Dipper is what astronomers call an *asterism* — a grouping of stars that is not an official constellation. The constellation is Ursa Major, the Great Bear, which is much larger consisting of additional fainter stars to the right and below. These form the head and the legs of the bear. The handle of the Plough or Dipper is the tail of the Bear (Figure 18).

Then there is another constellation of seven stars, Ursa Minor, the Little Bear, where the Pole Star is the last star of the tail (Figure 19).

Figure 19. Circumpolar stars circle around Polaris, the Pole Star, and never set. The higher the latitude, the more stars are circumpolar. The circle shown is for latitude 50°N.

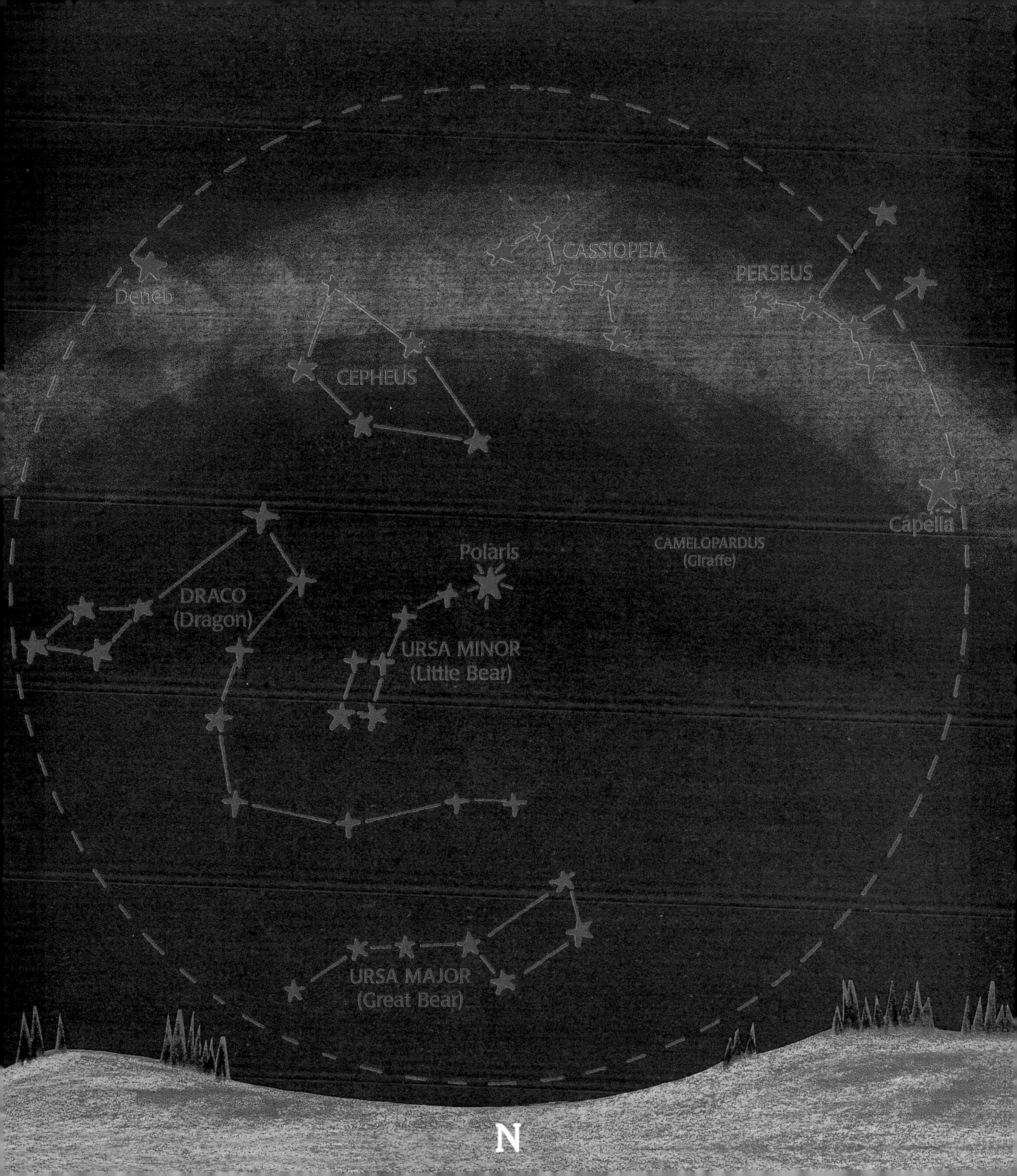

Deneb

CASSIOPEIA

PERSEUS

CEPHEUS

Polaris

DRACO
(Dragon)

CAMELOPARDUS
(Giraffe)

Capella

URSA MINOR
(Little Bear)

URSA MAJOR
(Great Bear)

N

The two furthest stars (closest to Ursa Major) are a little brighter than the others.

Opposite the tail of the Great Bear, about the same distance again past the Pole Star, is the W-shaped constellation of Cassiopeia. In the evenings of December it is high overhead, while during June evenings it is low in the north.

Between Ursa Major and Ursa Minor is a chain of stars snaking its way around Ursa Minor and then turning sharply away towards a diamond of stars. This is Draco, the Dragon, with a long tail and body and a diamond-shaped head. The furthest star of the diamond is officially part of Hercules. (Only in 1922 was the sky carefully divided into 88 internationally recognised constellations ensuring that a bright star was never on the boundary between them.)

Between the sharp turn of the Dragon and Cassiopeia is the constellation of Cepheus. Again it is not easy to recognise a figure in these five stars, which appear a bit like a house with a steep roof. King Cepheus and Queen Cassiopeia were mentioned earlier in the autumn constellations.

The space between Cassiopeia and Ursa Major has no bright stars. Only in a very clear sky can the constellation of Camelopardus, the Giraffe, be recognised. This is not one of the classic Greek constellations: it was named by the Dutch astronomer Plancius in 1613.

Depending on latitude, part of Perseus and Capella in Auriga, the Charioteer, may be circumpolar.

Finding your way around the constellations

Apart from finding the Pole Star from the Plough or Big Dipper, there are some other pointers to help us find some of the constellations (Figure 20).

- Following the pointers from the Plough through the Pole Star shows the constellation of Cepheus, and beyond that the square of Pegasus.
- Extending that line in the other direction, we find Leo, the Lion.
- Extending a line from the other two stars of the rectangle of the Plough we find Vega in Lyra and beyond that Atair in Aquila, the Eagle.
- Following the curve of the handle of the Big Dipper (or Plough) we find Arcturus in Boötes and beyond that Spica in Virgo.
- Extending a line along the top two stars of the Plough, away from the handle, we find Auriga, the Charioteer.
- The diagonal through the rectangle of the Plough points to Pollux in Gemini, the Twins.
- Taking a line from the handle of the Big Dipper through the Pole Star we find Cassiopeia.
- Beyond that is the curve of stars of Pegasus, Andromeda and Perseus.
- Extend that curve beyond Perseus and we find Capella in Auriga, the Charioteer.
- The line from Deneb in Cygnus, the Swan, through Atair in Aquila, the Eagle, points to Sagittarius, the Archer.

Figure 20. Finding your way around the northern skies.

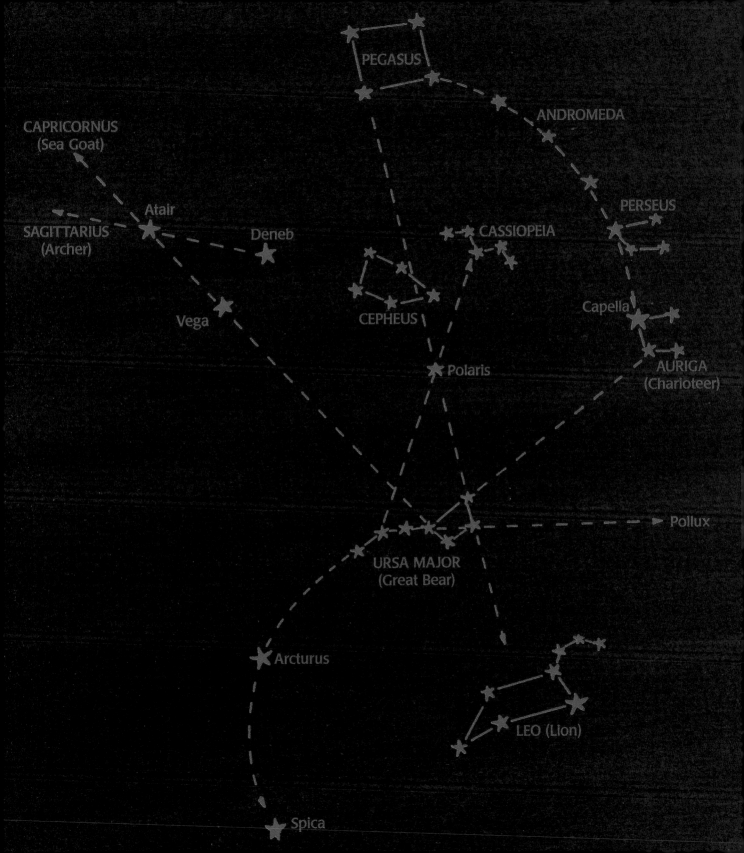

- The line from Vega in Lyra through Atair points to Capricornus, the Sea Goat.
- Extending the western side of the square of Pegasus we find Fomalhaut in Piscis Australis, the Southern Fish.

To get an overview of the stars at a particular time of night and time of year, a planisphere is very useful, if not essential. You can buy one, or make one as described on pages 55ff.

The Northern Hemisphere Constellations: Summary

- At each time of year a particular configuration of constellations can be seen. There is the Winter Hexagon, the Spring Triangle, the Summer Triangle and the Autumn Square. During the curse of the year new constellations appear in the mornings, some months later they culminate at midnight and are visible all night, and again some months later they disappear in the evening twilight.
- In the northern sky the same stars are visible all night, but in different positions at different times of year.
- The appearance of different stars at different times of the year can be explained by the annual revolution of the Earth around the Sun.

Southern hemisphere

In populated parts of the southern hemisphere all the constellations discussed above, except the circumpolar ones, are visible, though the other way up. Visitors to the opposite hemisphere, familiar with the stars at home, find they all appear upside down.

In addition new stars are visible. Crux, the Southern Cross is the smallest constellation with four fairly equally bright stars. The longest axis of the cross, extended about four times, points to the celestial south pole. There are two nearby bright stars, the Pointers (actually part of the constellation of Centaur). Extend the perpendicular of the line connecting the two, and where it crosses the line of the Southern Cross is the south pole (Figure 21).

Apart from the dark area around the south pole, the southern sky is full of stars. The Centaur (with the Pointers) encloses the Southern Cross. It is one of the classical Greek constellations, depicting the mythical creature that is half man, half horse (not to be confused with Sagittarius, a similar mythical creature). The pointer furthest from the Cross, alpha Centauri or Rigel Kentaurus, is the star that is closest to the Earth. Near the Centaur are Vela (the Sails), and Carina (the Keel) of the Greek ship Argo that Jason and the Argonauts sailed to retrieve the Golden Fleece. Canopus, the bright star at the end of the Keel is the second brightest star in the skies (Sirius is the brightest). The Milky Way stretches between these constellations, and is particularly bright in this region. On moonless clear nights it practically glows (Figure 22).

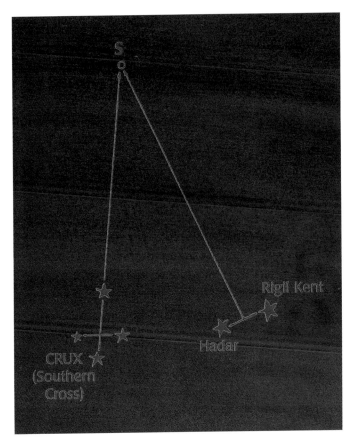

Figure 21. Finding the south pole using the Southern Cross and the Pointers.

Don't miss the two Magellanic Clouds that look like separated eddies of the Milky Way, in the constellations of Dorado and Tucana. Like the Andromeda Galaxy in the northern skies, they are galaxies. These three are the only galaxies that can be seen with the naked eye.

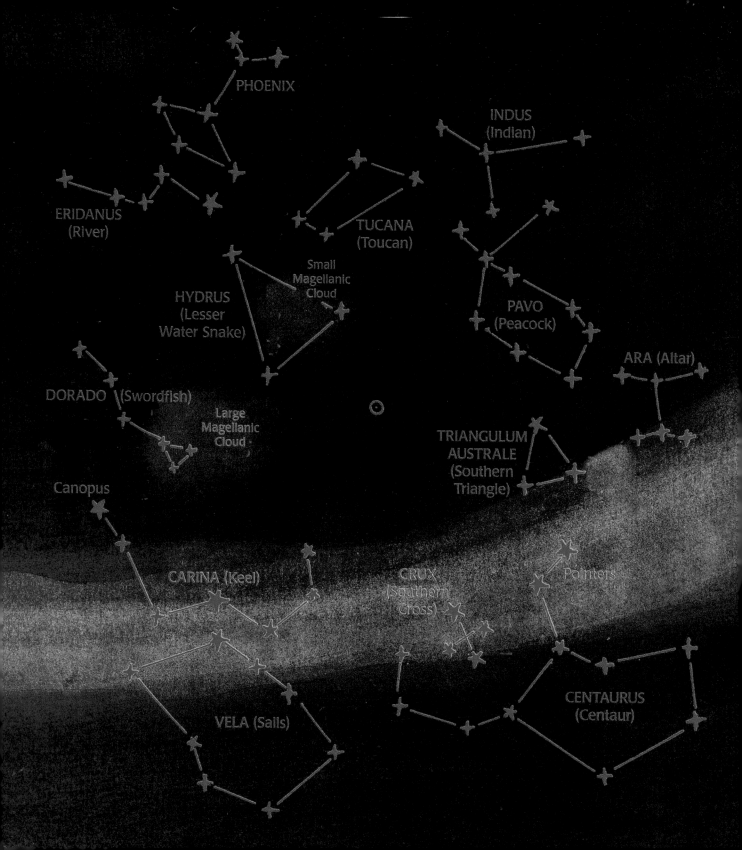

PHOENIX

INDUS
(Indian)

ERIDANUS
(River)

TUCANA
(Toucan)

Small
Magellanic
Cloud

HYDRUS
(Lesser
Water Snake)

PAVO
(Peacock)

ARA (Altar)

DORADO \(Swordfish)

Large
Magellanic
Cloud

TRIANGULUM
AUSTRALE
(Southern
Triangle)

Canopus

Pointers

CARINA (Keel)

CRUX
(Southern
Cross)

VELA (Sails)

CENTAURUS
(Centaur)

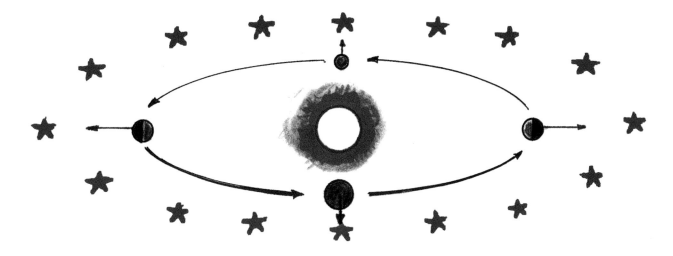

Figure 23. The annual revolution of the Earth around the Sun. The stars are much further away. The arrows point in the direction of stars visible at midnight.

The view from space

If look at the revolution of the Earth around the Sun — the Copernican viewpoint — we can see that the night-side of the Earth (the side opposite the Sun) points in different directions at different times of year. The arrows in Figure 23 show the different directions. The movement of the Earth is anticlockwise (seen from the north), corresponding to the anticlockwise movement of the Sun through the zodiac. The daily anticlockwise rotation of the Earth shows new stars becoming visible before dawn.

Figure 22. Constellations around the south celestial pole.

Mythology and star names

We have sometimes mentioned the mythology connected with a constellation. To do this thoroughly would go beyond the scope of this book, and can be found in other books. The brighter constellations of the northern hemisphere were usually named by the ancient Greeks and the mythology connected with them is Greek. However, the names by which they are known internationally are almost all Latin, though some are also known by their English names. To confuse matters further, the names of individual stars are often of Arabic origin and have been Latinised.

4. The Zodiac

There is a band of stars which is particularly important. This band, the zodiac, is significant because it is against its background that the Sun, Moon and planets move. Of course the zodiac also has a daily movement across the sky and we have described each of its twelve constellations in the seasonal descriptions (in the previous chapter).

Figure 24 shows the whole zodiac with the constellations on the outside, their symbols on the inside. In the northern hemisphere they appear as depicted in the top half of the circle (turning it for the right time of day or year), while in the southern hemisphere the constellations appear the other way round, as shown in the lower half.

Traditionally the zodiac begins with Aries. The little constellation Triangulum, the elongated Triangle, points to the beginning. Sun, Moon and planets move around the zodiac in an anticlockwise direction; in the southern hemisphere this movement appears clockwise.

The entire zodiac can never be seen all at once: at any time only half can be seen, simply because we can only see half the sky at any time, as the other half is below the horizon.

The zodiac has a higher and a lower part.

This is shown in Figures 25 and 26 for the northern hemisphere, and in Figures 29 and 30 for the southern hemisphere. We shall refer to these positions as high and low. The zodiac changes between these two positions every day. The difference in height is twice 23½°, that is, 47°.

It takes 12 hours from the highest to the lowest position. Between these two the zodiac tilts to the left or the right, as shown in Figures 27 and 28 for the northern hemisphere, and in Figures 31 and 32 for the southern hemisphere. A complete rotation of the zodiac takes almost 24 hours, making a kind of pulsating 'wobble'.

Figure 24. The zodiac. The constellations are on the outside, their symbols on the inside. In the northern hemisphere they appear as depicted in the top half of the circle (turning it for the right time of day or year), while in the southern hemisphere the constellations appear the other way round, as shown in the lower half.

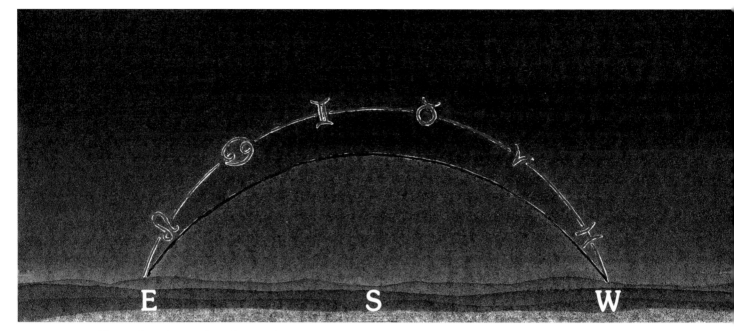

Figure 25. Northern hemisphere high zodiac at midnight in December, evening in March, midday in June and morning in September.

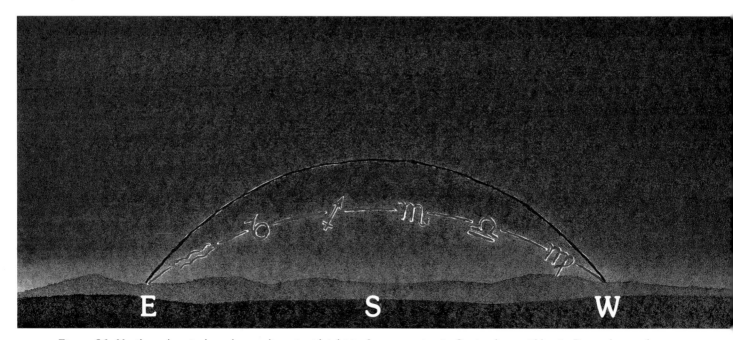

Figure 26. Northern hemisphere low zodiac at midnight in June, evening in September, midday in December and morning in March.

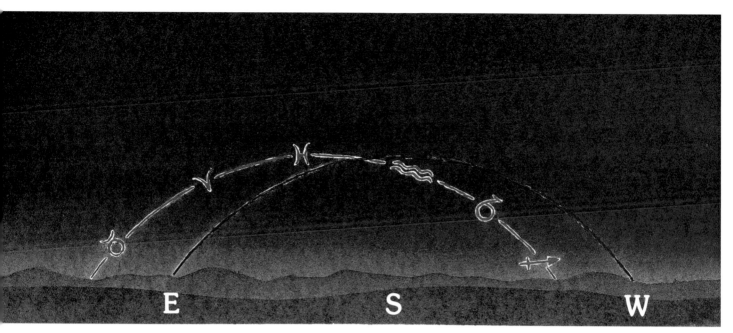

Figure 27. Northern hemisphere tilted zodiac at midnight in September, evening in December, midday in March and morning in June.

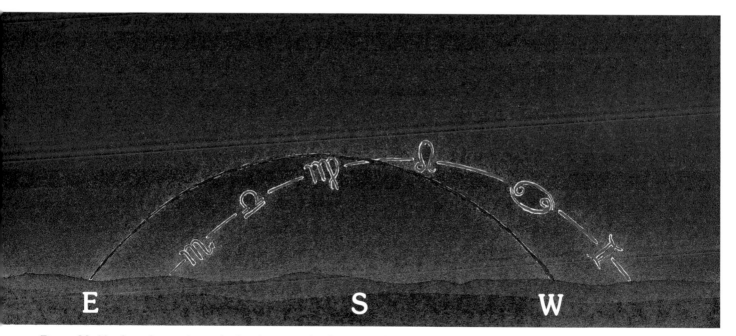

Figure 28. Northern hemisphere tilted zodiac at midnight in March, evening in June, midday in September and morning in December.

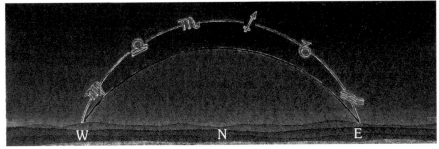

Figure 29. Southern hemisphere high zodiac at midnight in June, evening in September, midday in December and morning in March.

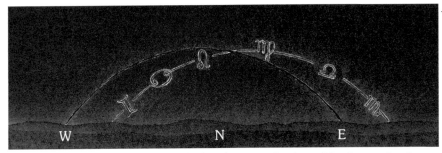

Figure 30. Southern hemisphere low zodiac at midnight in December, evening in March, midday in June and morning in September.

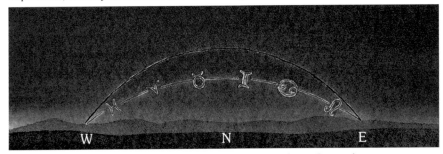

Figure 31. Southern hemisphere tilted zodiac at midnight in March, evening in June, midday in September and morning in December.

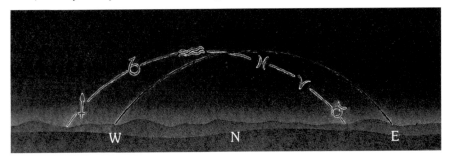

Figure 32. Southern hemisphere tilted zodiac at midnight in September, evening in December, midday in March and morning in June.

The central line of the zodiac (shown in yellow) is called the *ecliptic*. It is the line along which the centre of the Sun appears to move. Regulus in Leo, Spica in Virgo, and the foot of the higher (northerly) Twin of Gemini lie almost exactly on the ecliptic. Aldebaran in Taurus and Antares in Scorpius lie a little below it (or above it when seen in the southern hemisphere).

The irregular sized constellations of the zodiac show a certain pattern: after two large constellations there is a small one.

In the northern hemisphere the high constellations of the zodiac, Taurus and Gemini, rise to the north of east and set to the north of west, moving in high arcs across the sky; the low constellations, Scorpius and Sagittarius, rise and set to the south of east and west and move in low arcs. In the southern hemisphere, the high constellations, Scorpius and Sagittarius, rise and set to the south of east and west, moving in high arcs across the northern sky, while the low constellations, Taurus and Gemini, rise and set to the north of east and west, moving in low arcs across the sky.

The *celestial equator* (marked in red) divides the upper half of the ecliptic from the lower. These parts are also called the northern and southern halves, rather than the higher and lower, to avoid confusion in different parts of the world. The northern half stretches from Pisces through Taurus to Leo, and the southern part from Virgo through Sagittarius to Aquarius. The *equinox points* (where ecliptic and equator cross) are between Aquarius and Pisces (spring equinox in the northern hemisphere) and between Leo and Virgo (autumnal equinox in the northern hemisphere). In the southern hemisphere they are sometimes given the opposite names to match the season there. These two points are the crossing points of the celestial equator and the ecliptic. We shall look at them in more detail in the chapter on the Sun.

There are four bright stars at cardinal points of the zodiac which the ancient Persians called the royal stars. They are Aldebaran in Taurus, Regulus in Leo, Antares in Scorpius and Fomalhaut south of Aquarius. In old paintings these creatures can be recognised symbolising the four Evangelists — Luke as the Bull (Taurus), Mark as the Lion, John as the Eagle (in place of the Scorpion), and Matthew as the angel (or human being as Water-Carrier).

The Zodiac: Summary

- The zodiac consists of twelve constellations encircling the entire heavens. At any time only half the zodiac can be seen.
- The Sun moves along the central line of the zodiac, the ecliptic. The Moon and planets also move through the zodiac, though not exactly along the ecliptic.
- Half the zodiac is above the celestial equator, half below.
- During the course of the day the orientation changes, displaying a kind of pulsing 'wobble'.

5. Other Aspects of the Stars

The Milky Way

Like the zodiac, the Milky Way is a band right across the skies. It can be seen on clear nights, but city lights will drown out its subtle luminance. The band varies in width and in brightness, showing a few 'dark patches'. The stars of the Milky Way are part of the fixed stars and show a daily movement and changes in the time of year.

It follows a course between Sirius and Procyon, between Orion and Gemini, close by Capella and through the constellations of Perseus and Cassiopeia, Cygnus and Aquila to Sagittarius and Scorpius, and continues through Centaur, Southern Cross, Carina, Vela and Puppa (the Keel, Sail and Stern) in the southern celestial hemisphere.

Looking at the Milky Way through a telescope you can see it consists of thousands of tiny stars close together. There are so many of these faint stars that *every person on Earth could have their own.*

Sidereal time

As well as our *civil time* which we use in everyday life, astronomers use *sidereal time* which is not dependent on the Sun's course, but is linked to the movement of the stars. Every observatory has a clock showing sidereal time which indicates the position of the stars regardless of time of day or time of year. (The stars are above the horizon during the day, even if we cannot see them.) Zero hours sidereal time is when the spring equinox point culminates in the meridian. At that point the zodiac is as shown in Figures 27 (northern hemisphere) or Figure 32 (southern hemisphere). However, as the spring equinox is not a visible point, in the northern hemisphere we can take our orientation from the Plough or Big Dipper. It is low in the sky at 0^h (0 hours) sidereal time, at 6^h it is to the right of Polaris (the Pole Star), at 12^h high in the sky, and at 18^h to the left of Polaris. In the southern hemisphere we can use the Southern Cross as a reference: at 0^h it is low in the south, at 6^h it is to the left, at 12^h it is high and at 18^h it is to the right.

Making a planisphere

To make a planisphere suitable for latitude 50°N you can copy the two images in Figures 33 and 34. If possible enlarge them. Stick the copies onto stiff card and cut them into circles. Cut out the oval hole of the top card. Stick a clear foil over the back of the hole and cut off any surplus foil.

With a sharp tool make a hole in the centre of the each disc. The lower card (the star map) has the centre marked; on the upper card the centre can be found by marking the crossing point of lines N–S and 6–18. Lay the top card over the star map and fasten them together with a split pin (Figure 35). Now the planisphere is ready to use (Figure 36).

Directions for use

To make the planisphere show the stars visible, turn it so that the time of day and date match on the outside rim. The times are marked in a 24-hour clock, so for pm times add 12 (10 pm is 22 hours). The oval window then shows the stars which are visible above the horizon. Turn the planisphere for the direction in which you are facing, so looking north you should have N at the bottom.

It can be used between latitudes 45° and 55°N. If you are further south, more stars will be visible over the southern horizon than are shown, and in the north you will not be able to see stars close to the edge of the oval window.

You can buy planispheres for various latitudes from 51.5°N to 23.5°N and for southern hemisphere there is one for 35°S. Planispheres don't work well in equatorial regions.

Fixed stars

It should be mentioned that, strictly speaking, the fixed stars are not 'fixed'. Over long periods of time they change position, and the shape of constellations changes. However, during a human lifetime they are, for all practical purposes, 'fixed' and certainly a planisphere will be true for much longer than it lasts.

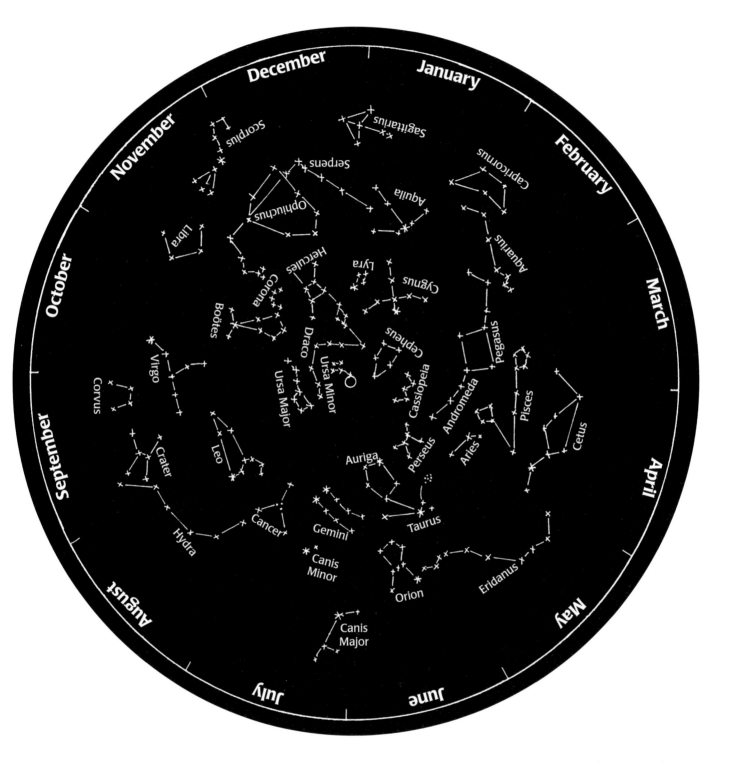

Figure 33. The star map, lower part of the planisphere.

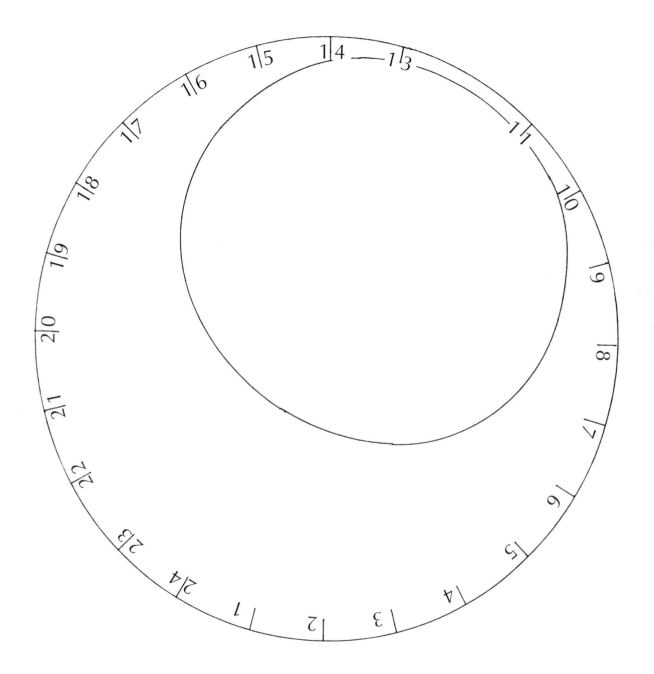

Figure 34. The upper part of the planisphere.

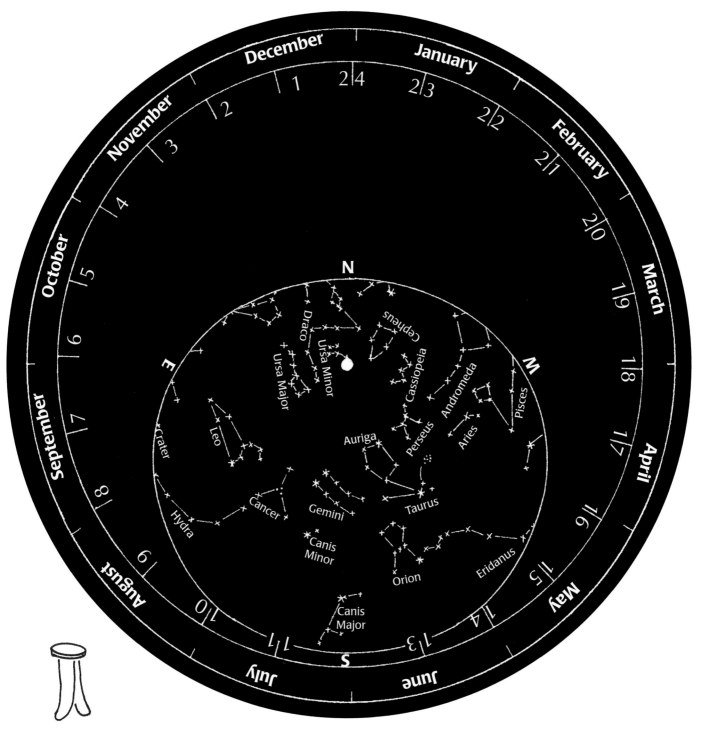

Figure 35. A split pin.

Figure 36. The finished planisphere, set for midnight, December 31.

PART II

THE SUN

6. The Sun's Apparent Movement

Until now we have looked at the stars which maintain their positions in relation to one another as they move across the sky in their daily rising and setting. We shall now turn to the Sun. The Sun is not only responsible for daylight and warmth, but also for plant growth, the colours in nature and indirectly almost all our energy. Astronomers call the Sun our nearest star.

The daily movement

At dawn the stars of the sky fade away in the growing light. The colours in the eastern sky announce the arrival of the Sun which then rises somewhere in the east. It does not rise straight up (except at the equator), but in the northern hemisphere moves diagonally to the right towards south (Figure 37). Because the Sun is so bright it is difficult to follow it, though we can look at a shadow and see the opposite movement. At noon it is high in the south, culminating on the meridian, before sinking gradually towards the west (Figure 38). This is followed by dusk and soon the first stars can be seen. At night, the Sun is below the horizon in the north.

In the southern hemisphere the Sun rises in the east, moves diagonally to the left, and then culminates in the north before sinking further left in the west. At night it passes below the horizon in the south. So in the southern hemisphere the Sun moves in the opposite direction to the north.

While stars culminate at all times of the day and night, the Sun culminates every day at noon, and we calculate our civil (everyday) time from this.

Every day the Sun moves across the sky, rising in the east and setting in the west. The shadow moves in the opposite direction: in the morning there is a long shadow pointing west, at noon a shorter shadow in the north (in the

Figure 37. Sunrise in northern temperate latitudes (40°N shown). The Sun rises at an angle to the right.

Figure 38. Sunset northern temperate latitudes (48°N shown). The Sun sets at an angle to the right.

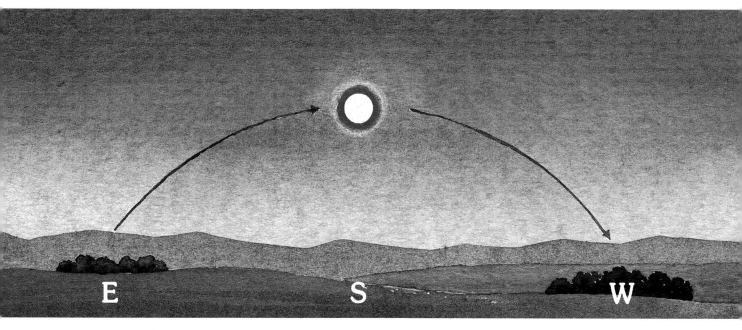

Figure 39. The arc of the Sun at the spring or autumn equinox in temperate northern latitudes. Sunrise is at 6 am exactly in the east, sunset exactly in the west at 6 pm.

south in the southern hemisphere), and in the evening a long shadow pointing east.

Depending on latitude, the Sun rises faster or more slowly. At the equator where it rises vertically it takes 2 minutes from the first appearance of the top of the Sun's disc until it is fully above the horizon. At latitude 40° (north or south) it takes just over 2½ minutes, but at latitude 60° it takes 4 minutes.

Twilight

The sky is still light while the Sun is less than 6° below the horizon. This time is called *civil twilight,* and it for instance determines when street lighting goes off and comes on. When the Sun is 12° below the horizon, the horizon is still quite visible, bright stars can be seen (in clear weather) but it is not possible to see objects on the ground clearly. This time is called *nautical twilight.* Astronomers are more particular and want a completely dark sky. This occurs at *astronomical twilight,* when the Sun is 18° below the horizon.

As the angle of setting is steeper nearer the equator, the time of twilight is shorter, while at high latitudes where the Sun rises and sets at a much shallower angle, twilight lasts much longer. For those not used to it, it can be disconcerting how quickly it becomes dark at lower latitudes (just as those from equatorial regions find the long twilight unusual).

In high latitudes at midsummer the Sun at midnight is not far below the horizon. In Stockholm, Sweden or Anchorage, Alaska (at 60°N latitude) the Sun is only 7° below the horizon, so that it is barely dark.

The Sun's annual course

Unlike the stars, the Sun changes its path across the sky during the course of the year, having higher and lower arcs.

Northern temperate latitudes

In spring and autumn, on March 21 and September 23, the Sun rises exactly in the east and sets exactly in the west wherever we are in the world. Sunrise is at 6 am and sunset at 6 pm (the exact time may be a later or earlier for both, depending on the time zone and possible daylight saving time). The day is exactly 12 hours long and the night is exactly 12 hours, which is why this time of year is called the *equinox* (Latin for equal night). This is shown in Figure 39.

In winter, at the solstice on December 21, the Sun rises later (about 8 am at latitude 50°N, 7:15 am at 40°N), and to the south of east. It culminates much lower in the south, casting a relatively long shadow. It sets earlier (about 4 pm at 50°N, 4:45 pm at 40°N) and to the south of west (Figure 40). It takes less than 12 hours to travel along the low arc (only 8 hours at 50°N, 9½ hours at 40°N), and spends a greater time below the horizon.

In summer, at the solstice on June 21, the Sun rises earlier (about 4 am at latitude 50°N, 4:30 am at 40°N), and to the north of east. It culminates much higher in the south, casting a relatively short shadow, and then sets later (about 8 pm at latitude 50°N, 7:30 at 40°N) and to the north of west (Figure 41). It takes more than 12 hours to travel along the higher arc (about 15 hours at 40°N), and spends a shorter time below the horizon. As we go further north the points of sunrise and sunset move more towards north from east and west during summer. This means that the Sun describes greater arcs across the sky, and that the days are longer. At 50°N the longest day is 16 hours, while at 60°N (Stockholm or Anchorage) the Sun is above the horizon for over 19 hours.

During the course of the year, the point of sunrise moves, swinging to the north and south of exact east (Figure 42). At latitude 50° this movement is about 36° on either side of east, so 72° in total. There is a corresponding movement of the point of sunset.

The Arctic Circle

If we keep travelling northwards at midsummer, we eventually reach a place where the points of sunrise and sunset have moved so far north that they coincide exactly in the north, and at midnight the Sun is on the horizon. In other words the Sun has become circumpolar, and is called the *midnight Sun*. This happens at a latitude of 66½°, a line called the *Arctic Circle*. It runs through northern Scandinavia, Siberia, Alaska, Canada, Greenland and touches the north of Iceland. Relatively few people live that far north — there are a few towns in Norway (Tromsø, Bodø), Sweden (Kiruna), Finland (Rovaniemi), Russia (Murmansk). In North America and Greenland there are only a few settlements.

However, if we travel there at midwinter, then the points of sunrise and sunset will have moved so far south that they coincide exactly in the south, and at noon half the Sun briefly appears on the horizon (Figure 43). Dawn and

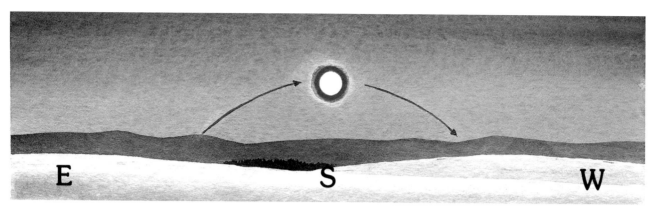

Figure 40. The low arc of the winter Sun in temperate northern latitudes.

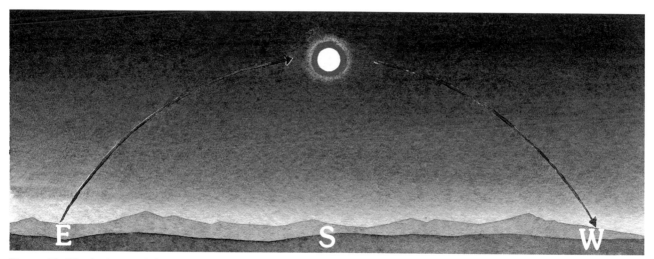

Figure 41. The high arc of the summer Sun in temperate northern latitudes.

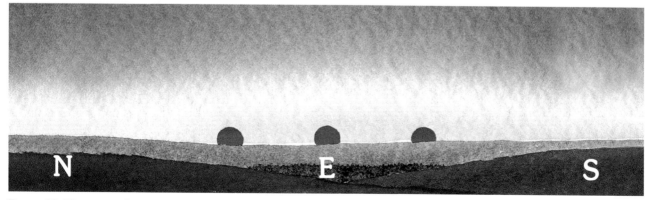

Figure 42. The point of sunrise in temperate northern latitudes. Only at the equinoxes does the Sun rise exactly in the east.

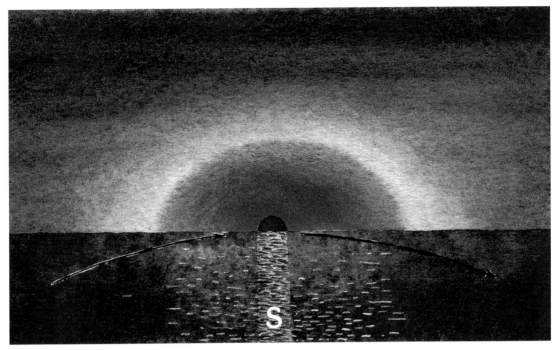

Figure 43. Noon on December 21 looking south from the Arctic Circle

dusk precede the brief sunrise/sunset at noon, but last for about twenty hours.

North of the Arctic Circle

As we travel north of the Arctic Circle, the number of nights with midnight Sun, and correspondingly the number of winter days without Sun, increase.

At the North Cape of Scandinavia there are six weeks of midnight Sun every summer (Figure 44). In winter for six weeks around noon the sky lightens to twilight, but the Sun never rises. Further north, in Spitsbergen (80°N) it does not even get light at noon in winter.

North Pole

At the North Pole itself we can only see the Sun when it is in the northern part of the zodiac, that is, between March 21 and September 23. During this half of the year, the Sun never sets. The polar 'day' last six months, and similarly the polar 'night' lasts six months. Dawn twilight begins some weeks before the spring equinox when the Sun rises, taking about 30 hours for the full disc of the Sun to rise above the horizon. During that time the Sun travels more than once around the whole horizon (Figure 45). Over the next three months the Sun gradually reaches its highest position of $23\frac{1}{2}°$ above the horizon at the summer solstice. Then the whole sequence is repeated in reverse — a long sunset on September 23,

Figure 44. The midnight Sun from the Lofoten Islands, Norway (68°N). Five photographs taken at hourly intervals, with midnight in the centre.

several weeks of twilight, and then complete darkness in winter.

During the summer, then, you cannot estimate the time of day from the height of the Sun in the sky. In fact, the only way to keep track of the time in the absence of a watch is to divide a circle into 24 and set up a post in the centre of it to observe the Sun's shadow.

The situation at the South Pole is similar, though the seasons are, of course, the opposite.

Figure 45. Sunrise at the North Pole takes place once a year at the spring equinox, and lasts about 30 hours.

The tropics

Going closer to the equator from the north, we reach a place where, at the summer solstice, the Sun is directly overhead at noon, barely casting any shadow and could be seen reflected in a deep well. The latitude at which the Sun is overhead on one day in the year (the June solstice) is called the *northern tropic* (from the Greek *tropos* to turn) or *Tropic of Cancer.* That latitude is 23½°N, and runs through northern Mexico, the Caribbean (just north of Havana, Cuba); the Sahara Desert and southern Egypt, Saudi Arabia; Gujarat and Madhya Pradesh in India, Dhaka in Bangladesh, through southern China and Taiwan and across the Pacific Ocean just north of the Hawaiian Islands.

In the days immediately before and after the summer solstice the Sun does not culminate overhead, but a little to the south of the zenith. (At the winter solstice at midnight the Sun will be exactly opposite in the nadir, the point directly below us.)

Continuing south of the northern tropic the Sun will be in the zenith twice a year. Close to the tropic these two days will be shortly before and after the summer solstice. As we move closer to the equator the interval between them increases.

The equator

If we continue southwards as far as the equator, the celestial equator goes through the east point of the horizon, the zenith overhead, the west point and below us through the nadir.

The Sun is exactly in the zenith at noon on March 21 and September 23 (the equinoxes). On these days it rises and sets exactly in east and west. During the course of the year its rising, culmination and setting points move 23½° to the north (in the northern hemisphere's summer) and to the south (in northern winter). The difference between rising and setting points has not diminished to nothing, but to ± 23½°, that is 47°.

Every day on the equator is equally long — 12 hours daylight and 12 hours night. The rhythm of the year is barely discernible at the equator; only the seasonal changes of rainy and dry times are noticeable.

Southern tropical latitudes

As we travel further south of the equator, all the changes we have seen in the north repeat in reverse order. A little south of the equator not much changes — the luminaries rise in an easterly direction and set in westerly parts, but the stars' and the Sun's paths are inclined to the north. The Pole Star can no longer be seen, but instead the south celestial pole is above the horizon. Unfortunately there is no prominent star there.

The two days of the year when the noon Sun is in the zenith come closer together, until at the southern tropic, at latitude 23½°S, they coincide on December 21.

The northern and southern tropics are also called the Tropic of Cancer and Tropic of Capricorn respectively. However, as the spring equinox (and thus also the summer and winter solstice) has moved in the zodiac, the actual constellation the Sun is in at the time of the solstices no longer coincides with Cancer and Capricorn (see pages 80f).

The further south we travel the more clearly the seasons appear, but the opposite way round from the northern hemisphere. When it is summer in the southern hemisphere, it is winter in the north, and spring and autumn are also at the opposite times of year. (Hence when we refer to the summer solstice or the spring equinox we need to make clear whether we mean the northern summer solstice — which is at the same time as the southern winter solstice — or the northern spring equinox, which is the southern autumn equinox.)

Southern temperate latitudes

Continuing south past the southern tropic the noon Sun is no longer found in the zenith, but is in the north. If we now face north and watch the Sun rise in the east, culminate in the north and set in the west, we see the movement is anticlockwise. This can be quite disconcerting for Europeans or North Americans visiting the southern hemisphere for the first time! (And of course Australians find it equally confusing when they come to the northern hemisphere and find the Sun moving the 'wrong way'.)

In winter, at the solstice on June 21, the Sun rises later (about 7:00 am at latitude 35°S), and to the north of east. It culminates much lower in the south, casting a relatively long shadow. It sets earlier (about 5:00 pm at 35°S) and to the north of west. It takes less than 12 hours to travel along the low arc (about 10 hours at 35°S), and spends a greater time below the horizon.

In summer, at the solstice on December 21, the Sun rises earlier (about 5 am at latitude 35°S), and to the south of east. It culminates much higher in the north, casting a relatively short shadow, setting later (about 7 pm at 35°S)

and to the south of west. It takes more than 12 hours to travel along the higher arc (about 14 hours at 35°S), and spends a greater time below the horizon.

The landmasses and populated parts of the southern hemisphere lie at lower latitudes than in the northern hemisphere. Santiago in Chile, Buenos Aires, Cape Town and Sydney all lie around latitude 35°S. Dunedin in New Zealand and Punta Arenas in Argentina, the southernmost sizeable cities, are at latitude 45°S and 53°S respectively.

The Antarctic Circle and the South Pole

Apart from a few research stations there are no towns or villages south of 55°S. The Antarctic Circle lies around the Antarctic continent. The aurora is visible from there, but almost no one sees the colourful spectacle. At the South Pole it is daytime for half a year in summer (from September to March) and night-time for half a year. Sun and stars circle from right to left, parallel to the horizon.

The Sun in the zodiac

We have already described the movement of the Sun, and separately the movement of the stars. We can summarise the two in Figure 46. The middle arc, from east to west, corresponds to the movement of the stars on the celestial equator, for instance the belt of Orion. The upper arc corresponds to the stars of the northern celestial hemisphere, and the lower arc corresponds to the movement of the stars in the southern celestial sphere. The Sun moves

Contrasts in different parts of the world

- In Central Europe, southern parts of Canada and the northern part of the United States there is a distinct daily rhythm and annual rhythm.
- At the poles of the Earth, the rhythm of the day disappears into the rhythm of the year. From there only half the stars can be seen.
- At the equator the rhythm of the year disappears, and every day is the same length. Twilight is short. All the stars can be seen from there.
- The Arctic and Antarctic Circles are the borderlines where the midnight Sun can be seen at least once a year.
- The northern and southern tropics are the borderlines where the noonday Sun reaches the zenith at least once a year.
- Everywhere on Earth has an equal share of daytime and night-time during the course of a year, but it is not the same everywhere. At the equator each 24 hours has an equal share, while at the poles the sharing is spread across the whole year.

$23\frac{1}{2}°$ to the north and south of the celestial equator.

From this we can deduce that the Sun does not always have the same position against the background of the fixed stars. It follows the arcs of different stars at different times of year. Each arc has a certain distance from the equator, and is parallel to it. Only the equator is a great circle — the others are all smaller (the higher arc looks bigger on the flat page in Figure 46, but is in fact smaller).

Looking at the different visible stars during the course of the year, it can be deduced that the Sun moves around the sky in a great circle, and does not simply swing to the north and south in one area of the sky. We can deduce by careful observation that the Sun moves through the band of stars that we know as the zodiac.

Every month (on average) the Sun moves into another constellation of the zodiac. This movement is in an easterly direction, that is, to the left in the northern hemisphere. Each day the Sun follows the daily movement of one of the constellations of the zodiac, though it is always of the one constellation that we cannot see, as the Sun's light outshines it. (The only time we can see the stars around the Sun is during an eclipse which we shall describe later.)

At midnight that part of the zodiac exactly opposite the Sun culminates, and thus an accurate picture of the Sun's movement in the opposite part of the zodiac can be calculated. The constellation that culminates at midnight is the one where the Sun was six months before and will be six months later.

The central line through the zodiac, the ecliptic, is the annual path of the centre of the

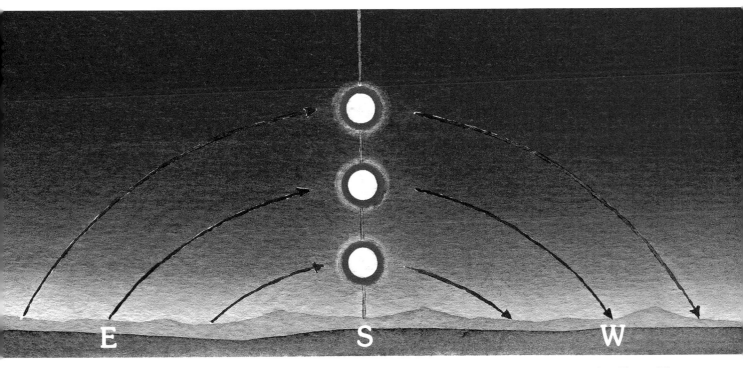

Figure 46. Daily arcs of the Sun at midsummer, the equinox and at midwinter in northern temperate latitudes. The middle arc is the same as the celestial equator.

Sun's disc. The ecliptic lies at an angle of 23½° to the celestial equator, so that half the zodiac is to the north of the equator, and half to the south.

In summary we can say that the Sun has two movements: the daily movement from east to west; and, against the background of the stars, a slow movement from west to east of about 1° every day completing the course through the zodiac in a year. While we cannot see these movements, they are the origin of the seasons.

For half the year the Sun is in the northern part of the zodiac and for the other half in the southern part. Twice a year there is a balance at the time of the equinox, on March 21, and September 23. These are the days the Sun crosses the celestial equator. The two points are called the spring (or vernal) equinox and the autumn equinox. March 21 is the spring equinox in the northern hemisphere, and the autumn equinox in the southern, and September 23 is the reverse. Conventionally, if the hemisphere is not specified, the northern hemisphere is assumed. As we have seen, on these two days the Sun rises exactly in the east and sets exactly in the west, no matter where we are on the earth.

The Sun moves in the highest and the lowest arc only once a year at the summer and the winter solstice, at July 21 and December 21

The Sun's movement against the stars

There are 12 constellations, so the Sun takes on average one month to move through each, though it varies between less than 3 weeks for the little constellations of Libra or Cancer and 6 weeks for the long constellation of Virgo.

As the Sun moves through the zodiac in the opposite direction of the daily movement, it lags behind the daily movement of the stars, taking about 4 minutes longer. That means that the stars complete their daily movement in about 23 hours and 56 minutes. This is the difference between sidereal time which astronomer use, and solar (the Sun's) time or civil time which we use for our everyday purposes.

and autumn, around the time of the equinoxes the daily change is greatest, while towards the solstices there is not much change from one day to the next.

As we saw, at the equator this change does not happen: each day and night is 12 hours long, but as the latitude increases (either to north or to south) the change becomes more noticeable, and extreme towards the Arctic and Antarctic circles.

The position of the spring equinox is between Aquarius and Pisces; the autumn equinox is between Leo and Virgo. The summer solstice is between Taurus and Gemini, and the winter solstice is between Scorpius and Sagittarius.

(or vice versa in the southern hemisphere). Again, conventionally, if the hemisphere is not specified, the northern hemisphere is assumed.

Between the winter and summer solstice the Sun is ascending; that is, the days become longer and the nights shorter, slowly at first, fastest around the equinox and then slowly again before the longest days. In the other half of the year the Sun is descending, again slowly at first and fastest at the equinox. In spring

The Sun's Apparent Movement: Summary

- The Sun has two movements in the sky. The daily movement is from east to west, like the stars move; this movement is comparatively fast, taking 24 hours to complete.
- The annual movement against the background of the stars is much slower, and goes in the opposite direction from west to east. This movement through the stars of the zodiac takes a full year.
- Because the zodiac is half above and half below the celestial equator, the Sun describes smaller and larger arcs across the sky.

- From the winter solstice to the summer solstice, the Sun ascends, and then in the other half of the year it descends. Midway in its ascending or descending, the Sun crosses the celestial equator at the equinoxes.
- The time of day is determined by the Sun's daily movement, while the seasons are determined by the Sun's progress through the zodiac.

7. The Sun's Movement Seen from Space

Earlier we looked at the rotation of the Earth, viewing it as if we were at a distant point in space, rather than standing on the ground. Let us look at the phenomena of the Sun from this viewpoint. The Sun always illuminates half the sphere of the Earth, that is, the *dayside.* The other half is in darkness and there is night. As the Earth rotates, day and night change in a 24-hour rhythm.

If we imagine ourselves at sunset, the Earth is rotating away from the Sun, and we are being moved into the dark side. We are delivered into the dark part of the earth through the Earth's rotation, and the Sun is no longer visible. It has 'set'. Similarly at sunrise we are transported (together with all our surroundings and the landscape) by the Earth's rotation from the dark side into the light side. As we are moved into

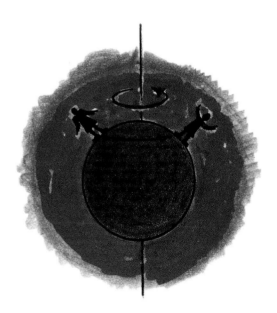

Figure 47. The Earth seen from the night side, with the Sun behind it. On the left is sunset and on the right is sunrise.

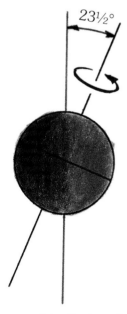

Figure 48. The axis of the Earth is tilted at an angle of 23½°. Imagine the Sun far away on the right.

Figure 49. Summer in the northern hemisphere. The Sun appears high in the sky at noon for an observer in northern latitudes. (You can imagine the opposite for an observer in the southern hemisphere). Imagine the Sun very far away.

the light part, we see the Sun rising and are soon dazzled by its brightness (Figure 47).

If the Earth did not rotate, half of it would be in continual darkness and half in continual light, and a narrow band between would be in eternal twilight. As the Earth does rotate, however, we all experience the 24-hour change from day to night and back again. Pedantically we could say that there is no such thing as sunrise and sunset, but it is a lot more convenient to call it that than 'the Earth has rotated so far that we are in darkness,' or 'the Earth has rotated to the point where sunlight replaces darkness.'

That is the *daily* change. To understand the *annual* changes, we have to take into account that the Earth's axis is inclined at 23½° to the axis of its annual revolution. Figure 48 makes it easier to understand.

The higher and longer arc of the Sun in northern summer is the result of the axis being tilted so that the northern hemisphere is towards the Sun (Figure 49). The noon position of the Sun is much higher in the sky, and the distance from noon to the shadow side is relatively long, resulting in a long afternoon and evening before dusk.

The low arc of the Sun in northern winter is the result of the axis being tilted so that the northern hemisphere is away from the Sun (Figure 50). The noon position of the Sun is low in the sky, and the distance from noon to the shadow side is relatively short, resulting in a short afternoon to dusk.

Figures 49 and 50 show the Sun once on the right and once on the left. How does the Sun get to the other side? According to Copernicus,

Figure 50. Winter in the northern hemisphere. The Sun appears low in the sky at noon for an observer in northern latitudes. (You can imagine the opposite for an observer in the southern hemisphere.) Imagine the Sun very far away.

it is not the Sun that moves, but the Earth that moves around the Sun. The seasons are explained by the Earth's revolution around the Sun, which takes one year. There is a difference of six months between the two illustrations. While the Earth moves around the Sun, the angle of the axis does not change, and the daily rotation is not affected.

Figure 51 shows the annual revolution of the Earth around the Sun. The diagram is in perspective so the Earth appears larger when close. The annual revolution around the Sun against the background of the fixed stars is the centrepiece of the Copernican theory. The path of the Earth lies in a plane (an imaginary flat surface). This plane is called the *ecliptic plane,* and the Sun also lies in this plane.

When the northern hemisphere of the Earth is tilted towards the Sun, it is summer in northern

parts (Figure 51, left). And at the opposite time of year (right) the northern hemisphere is titled away from the Sun. In spring and autumn the axis is also tilted, but neither away from or towards the Sun, and thus the day is neither lengthened nor shortened. It is the time of the equinoxes.

The Arctic and North Pole

Because of the 23½° tilt of the Earth to its annual path, the Sun illuminates the whole Arctic region (north of the Arctic circle) at the summer solstice (Figure 51 left). The 'highest' part of the Earth (as shown in the diagram) is at the borderline between light and shade, between the 'day' half and the 'night' half of the Earth. At that point the midnight Sun is just visible halfway above the horizon. Similarly at the winter solstice the whole Arctic region

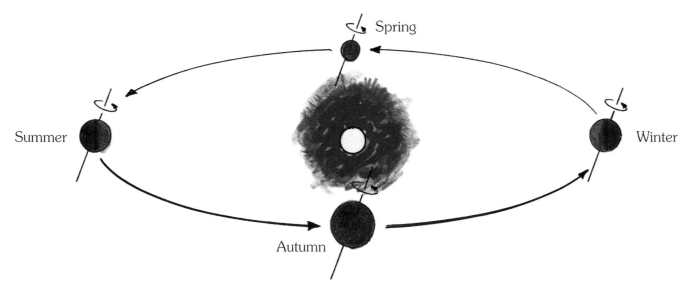

Figure 51. The Copernican system. The seasons arise out of the annual revolution of the Earth together with its angled axis of rotation. (The seasons marked are for northern hemisphere.)

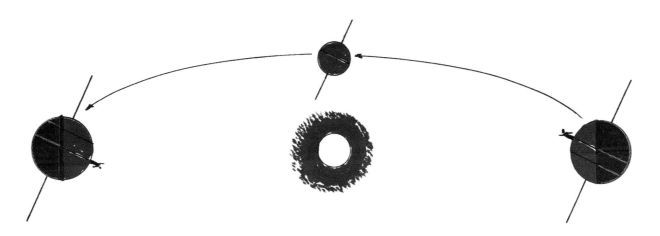

Figure 52. The two figures are standing on the equator pointing to the noon Sun. On the left the Sun is north of the zenith (northern summer solstice), and on the right the Sun is south of the zenith (northern winter solstice). The two extreme positions of the Earth are six months apart. At the back is the position of the Earth at the (northern) spring equinox.

(north of the Arctic circle) is in darkness (Figure 51 right). Only at the 'highest' part of the Earth, at the borderline between 'day' and 'night' will half the Sun appear at above the horizon briefly at noon.

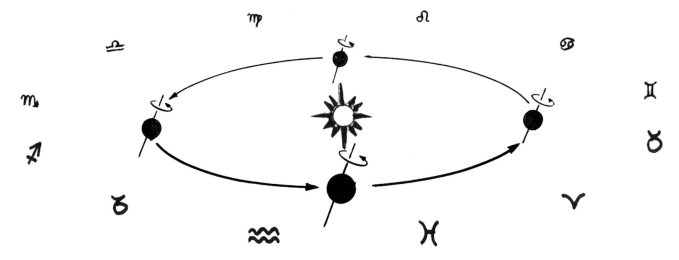

Figure 53. The Copernican system. The stars of the zodiac are in the ecliptic plane very far away.

The equator

Figure 52 shows the conditions at the equator. On the left is the Earth's position during the northern summer. At the solstice an observer at the equator, because of the Earth's tilt, sees the noonday Sun 23½° to the north of the zenith. The right hand side shows the equivalent situation at the northern winter solstice when the Sun is 23½° to the south of the zenith. From the northern or southern tropic the Sun can be seen in the zenith. At the equinox (behind), the Sun is in the zenith at noon.

At the equator, day and night are equally long at all times of year. No matter in what position the Earth is in its annual revolution around the Sun, the equator is always exactly half in the lit side and half in the shade side of the Earth. It is easy to see how going north or south of the equator, the length of day and night changes. The northern and southern tropic is the line of latitude at which the noon Sun is directly overhead at the solstice. Only an observer at the equator will always experience vertical rising and setting of stars and Sun, as the equator is perpendicular to the Earth's axis.

The zodiac

The seasons are connected to the Sun's position in the zodiac. To complete our overview, we have added the constellations of the zodiac in Figure 53. The stars should be imagined much further away from the Earth at varying distances.

Notice that the ecliptic, which we first saw as the central line of the zodiac, is now the plane of the Earth's orbit around the Sun (shown by the red arrows in the illustration). This is not contradictory: because we as observers are in that plane, it appears to us as a line, just like the horizon of the sea. The line of the zodiac and the plane of the Earth's orbit are one and the same, and are both called the ecliptic. The plane is usually shown horizontally and the axis of the Earth tilted. The ecliptic too has an axis perpendicular to it, with a north and a south pole. The *north ecliptic pole* is in the kink of Draco's tail, and the *south ecliptic pole* is in the constellation of Dorado. Seen from the Earth, the ecliptic poles have a daily rotation.

On the left of Figure 53 we see the Earth at the northern summer solstice. Seen from the Earth, the Sun is between the constellations of Taurus and Gemini. On the right the Earth is at the northern winter solstice when the Sun appears between Scorpius and Sagittarius. Between them are the spring (back) or autumn (front) equinox positions when the Sun appears between Aquarius and Pisces, and between Leo and Virgo respectively. The arrows show the direction of the Earth's annual movement, which means the Sun appears to move through the zodiac from right to left. (Figure 53 is drawn with north at the top; from a southern hemisphere perspective turn it upside down, and the annual movement is from left to right.)

We can see that the Copernican system does not contradict what we *see*, but helps us understand the phenomena. The $23\frac{1}{2}°$ inclination of the ecliptic to the celestial equator — the extension of the plane of the Earth's equator — can be seen from the angle of rotation to the plane of the ecliptic.

The intersection of the two planes means that half the ecliptic is to the north of the equator, and half to the south. We could equally say that half the equator is north and half is south of the ecliptic.

The angle of the Sun's path between sunrise and sunset can be seen not as the Sun being at an angle, but as our horizon being at an angle to the plane of the Earth's orbit.

Leap years

After 365 days (or 365 rotations) the Earth has not quite reached the same position in relation to the background of the stars, and thus to the seasons. It needs approximately another 6 hours — a quarter of a day — to reach the same position in relation to the year. That adds up to one day every four years, which we recognise as the extra day (February 29) that is added every four years. If the extra day was not added our calendar dates would creep back, and so January 1 would creep into autumn.

However, as the years is not quite 6 hours longer than 365 days, but a few minutes short of that, a further adjustment is made: every 100 years the leap day is omitted (as in 1900 or 2100), but every 400 years it is added (as in 2000). This adjustment was made in 1582 by Pope Gregory XIII and is called the Gregorian calendar reform. It was adopted at different times by different countries: Britain and its colonies (including what became the United States) adopted it in 1752.

We can also see the difference between sidereal (star) time and civil (Sun) time. If the Earth did not rotate, then we would see the stars standing still in the sky. During the course of the year the Earth would still be illuminated from all sides, but from the Earth we would see the Sun slowly rising once a year and slowly setting six months later. Looking at Figure 53 we can see that no matter where on the Earth we were, once a year the Sun would rise in the west and set in the east. Very slowly, the Sun would move the 'wrong way'. However, as the Earth *does* rotate once a day, the combination of these movements results in sidereal time being a little faster than civil (everyday) time.

If the Earth revolves around the Sun, it must surely be possible to notice a difference in the stars at opposite times of year. In fact, with very exact measurements some stars show a difference (Figure 54). This difference in angle is known as the *parallax* of stars. Only those stars that are closer to Sun and Earth show a measurable parallax, others are too far away to detect it. (The closest star has a parallax of 0.77 arcsecond, the equivalent angle of an object 2 cm in diameter, located 5.3 km away.) The German astronomer Friedrich Bessel first measured a star's parallax in 1838. This was another proof of the movement of the Earth around the Sun, for if the Sun moved around a stationary Earth there would be no parallax. (The other proof was Foucault's pendulum discussed on page 28.)

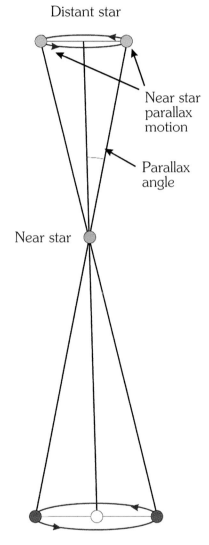

Figure 54. Measuring the parallax of stars.

The Sun's Movement Seen from Space: Summary

- The Copernican theory states that the rotating Earth revolves around the stationary Sun. The Earth moves in a plane, the ecliptic plane, and the centre of the Sun is also in this plane.
- Seen from the north, both the rotation and the revolution of the Earth are anticlockwise.
- The stars of the zodiac are roughly in the plane of the ecliptic, but very far away.
- The axis of the rotation of the Earth is inclined by $23\frac{1}{2}°$ to the axis of the ecliptic plane, and does not change significantly. This is the cause of the changing seasons.

8. Other Phenomena of the Sun

The Platonic Year

As well as the Sun's annual course through the zodiac there is another, much slower movement. Careful observation over long periods of time found that the equinox points are not stationary in the zodiac. They are always opposite each other, but they move very slowly around the zodiac from east to west, that is, in the other direction of the annual movement of the Sun. The time it takes for them to return to the same point is about 25 800 years, and this period is called the Great Year or Platonic Year. Thus a Platonic 'month' is about 2 150 years, which means that a full Platonic 'month' has not yet passed since the time of Christ.

At present the spring equinox is between Pisces and Aquarius, and the autumn equinox between Leo and Virgo. The spring equinox is slowly moving towards Aquarius. That is why there is talk of the coming Age of Aquarius. During the course of a human lifetime, the equinox points move about 1°, or twice the diameter of the Sun. You could say that a human life is as long as a Platonic 'day'.

When we looked at the seasons we simply ignored this. However, in another 2 150 years the equinox will have moved through Aquarius and come to Capricornus. And two thousand years ago the equinox moved into Pisces.

This movement of the equinoxes is also called the *precession* of the equinoxes. We can also imagine this movement as the zodiac shifting in the opposite direction. We saw earlier that the ecliptic (the central line of the zodiac) also has a north and south pole. The north pole is in the kink of Draco's tail, and the south pole is in the constellation of Dorado. However,

---

The Platonic Year

We breathe about 18 times a minute, that is, 1 060 times an hour, or about 25 920 times a day — about the same number as there are years in a Platonic Year — or one could say a year in a Platonic year is like a breath in a day. Our heart beats about 4 times per breath or about 72 times a minute — and the Sun moves 1° in about 72 years.

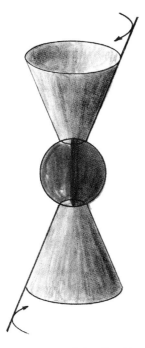

Figure 55. The precession of the Earth's axis during a Platonic Year makes a double cone with an angle of 2 x 23½°, or 47°.

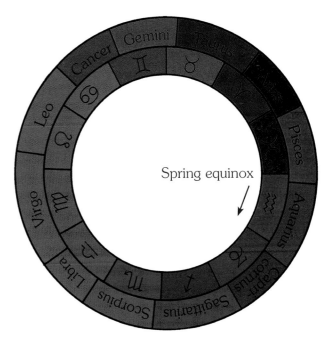

Figure 56. A comparison of the astronomical constellations which vary in size (outside circle) and the regular astrological signs (inside circle) at present.

as the equinox points move in the zodiac, the celestial equator moves, and thus the celestial poles also move around the ecliptic poles. That means that the Pole Star will not always mark the north pole of the sky.

Viewed from space, the precession of the equinoxes is like the 'wobble' of a spinning top. The axis of the Earth moves in a circle around the ecliptic pole. It is like a double cone whose point is in the centre of the Earth (Figure 55). After about 13 000 years, half a Platonic Year, the summer Sun will shine from the constellations where at present it is in winter.

Astrologers (those who study the influence of

stars on human life) have divided the zodiac into twelve equal parts of 30° each, the astrological *signs,* beginning with Aries on March 21. They ignore the precession of the equinoxes, and thus the astrological signs are shifted in relation to the astronomical *constellations.* As the constellations have varying sizes, the shift is roughly one constellation (Figure 56). This explains why people born in the sign of, say, Gemini find that the Sun is in the constellation of Taurus on their birthday. The astrological signs remain constant in relation to the seasons: in 13 000 years March 21 will still be at the beginning of the *sign* of Aries, while the Sun will be entering the *constellation* of Libra.

The Platonic Year: Summary

- Accurate observations over long periods of time show a slow shift of the equinox points in the zodiac. This movement takes about 25 800 years.
- This change is caused by the 'wobble' of the Earth's axis which also causes a movement of the north and south celestial poles around the ecliptic poles.
- In the distant past and the distant future the Pole Star no longer marks the celestial north pole.

The apparent diameter of the Sun

In order to view the edges of the Sun, astronomers use a *coronograph* — a specially designed telescope with a small disc to totally hide the light of the Sun itself and allow the flares and corona of the Sun as well as stars close to the Sun to be visible. (It must be emphasised that you should not try and make such a thing yourself — even a very brief look at part of the Sun through a telescope will permanently damage your eye. **Never look directly at the Sun!**)

The disc is made to be exactly the diameter of the Sun. However, during the course of a year the disc has to be changed for a slightly larger or slightly smaller one, as the diameter of the Sun changes a little during the year. Without an accurate measuring instrument this change is not noticeable, as it is only a variation of about 3.5%. The cause of this variation is that the distance of the Sun to the Earth does not remain constant, but is sometimes a little less, sometimes a little more.

A sundial

From the position of the Sun you can judge the time of day. A more precise estimate can be made if we let the shadow of a stick fall onto a surface marked with numbers. This is a sundial, and the stick or pointer is called a *gnomon*. Sundials can have a horizontal surface or a vertical one. It is important that the gnomon is angled correctly to be parallel to the Earth's axis, otherwise it will not show the time accurately through the year. That means if you make or buy a sundial, make sure it is correct for your latitude.

If you compare the time shown by a sundial with that shown on a clock, you will find noon on a sundial can be up to fifteen minutes earlier than clock noon (in May and November), or later (in February and August). From this you can deduce that the Sun moves a little faster or slower in the course of a year.

If you photographed the Sun every day at noon (clock time) from exactly the same location and combined all the photographs, you would get a narrow figure-of-eight (Figure 57). We are familiar with the change from a low winter Sun to a high summer Sun, but we also find a shift to the right (west, when the Sun passed the meridian before clock noon) or to the left (east, when the Sun only crosses the meridian after clock noon). Astronomers call the time shown by the Sun's position *apparent solar time* (where the Sun appears), and call clock time *mean solar time* (mean is another word for average). The difference between them is called the *equation of time,* and the figure-of-eight shows this.

As well as the changes in the Sun's apparent diameter, there are also a changes in the speed of the Sun's movement against the background of the stars. The two are connected in that the Sun in January is larger and moves faster, while in July it is smaller and slower. We shall look at this in greater detail below under 'Kepler's laws'.

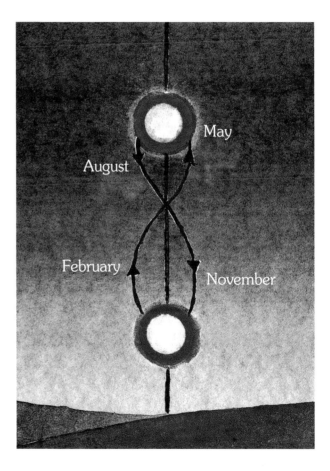

Figure 57. The noon position of the Sun during the year (northern hemisphere). The dark blue line is the meridian (above the south point of the horizon). The red line shows the actual position of the Sun at noon (clock time), or the apparent solar time compared to mean solar time. The difference is called the equation of time.

East-west differences

In our descriptions up to now we have shown how at different latitudes, that is, further north and further south, there are great differences in the movement of the stars, Sun and Moon. If we move east or west at the same latitude nothing much changes other than their rising and setting times as well as their time of culmination.

The change can be best experienced, of course, with the Sun. If we travel east, we are travelling against the daily movement. The Sun rises, culminates and sets earlier than where we started. During this journey days and nights become shorter. The opposite is true if we travel westwards where we follow the daily movement. Sunrise and sunset are earlier, the days and nights are longer.

The time difference is something we have to think of when phoning long distance. At noon in New York it is only 9 am in California, but already 5 pm in London. If we travel, we have to change our watches to the local time at our destination — travelling east it will be later, and travelling west it will be earlier. On longer journeys by air our body has to cope with this time change in a short time, and we often suffer from jetlag.

Whatever time we lose or gain we make up again on our return trip. However, if we travel completely round the Earth, this does not quite work. Magellan was the first person known to have organised a circumnavigation of the globe in 1519–22. Although he was killed in the Philippines, some of his crew managed to make it back to Spain. They kept a meticulous logbook and thought they had returned on a Sunday and were surprised to find that it was Monday. As they had sailed westwards, each day was a few minutes longer because they were at a different longitude. In the course of their three-year voyage this totalled 24 hours. At that time clocks were not sufficiently accurate to show the few minutes' difference in time each day.

In *Around the World in Eighty Days* by Jules Vernes, Phileas Fogg believes he has lost his bet to go around the world in 80 days, but Passepartout, his servant, finds out that only 79 days have passed in London, as they have been continually travelling eastwards, and so Fogg wins his bet.

Latitude can be determined anywhere very

The moving date line

There is another kind of date line that is not static, but moves. The line opposite the Sun, the midnight line, moves from east to west as the earth rotates and along it the date changes to the next day.

Figure 58. A road sign reminding of time zone change between Montana and northern Idaho, United States

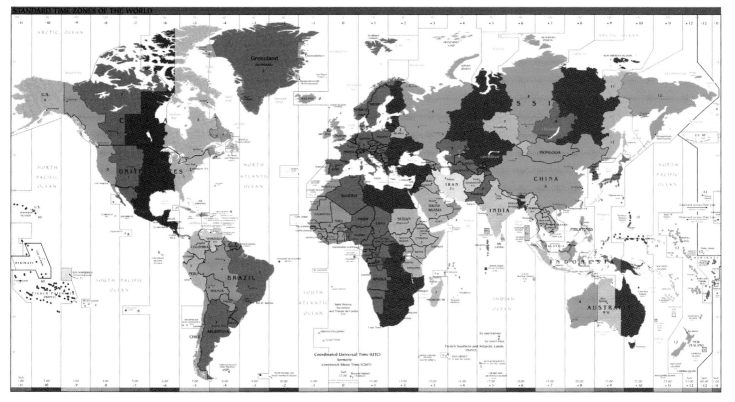

Figure 59. Time zones of the world

precisely by measuring the angle of the Pole Star above the horizon. However, this is not the case with longitude. In theory the simplest method to determine longitude is to take an accurate clock from a known location, and after travelling east or west determine the local time by the culmination of certain stars or the Sun, and compare this to the clock. But it was incredibly difficult to build a clock (or *chronometer* as accurate clocks are called) that was sufficiently precise for this, as pendulum clocks were pretty useless on a ship that tossed and turned on sea waves. In 1736 John Harrison succeeded in building a chronometer which was tested at sea, and spent the rest of his life perfecting it.

Time zones

The Earth rotates once in 24 hours through 360°, thus turning 15° per hour or 1° every 4 minutes. Thus travelling east or west our watches need to be adjusted by 4 minutes for *every* 1° of longitude we travel. At the equator this is about 111 km (69 miles), at latitude 45° this is 79 km (49 mi), and at latitude 60° only 56 km (35 mi).

Before railways, each town and city kept its own local time. With the demands of railway timetables most countries introduced a standard time in the late nineteenth century; later, in part due to air travel, standardised time zones one hour apart were introduced. So our local clock time (civil time) is no longer the true local time when our noonday Sun culminates.

For practical reasons the world is divided into time zones of 15°, each representing a one-hour time difference from the *prime meridian* of Greenwich (London). Almost all countries in the world use one or several of these time zones. (Parts of Australia, Newfoundland and India use a half-hour in-between time.) Some true local times vary from the zone time used there. For instance, longitude 0° passes through both France and Spain but these countries use Central European Time, which is one hour ahead. Similarly the whole of Alaska observes a time zone based on longitude 135°W (9 hours behind Greenwich), but Nome in the west (165°W) and the sparsely populated Aleutian Islands stretch well beyond that.

Places to the east of Greenwich are one or more hours ahead of Greenwich Mean Time, and places to the west of Greenwich are one or more hours behind. To avoid miscounting dates, the International Date Line runs roughly down longitude 180°, in the middle of the Pacific Ocean. Whenever we cross this line we not only have to change the time but also change the date, forward one day if travelling west, and back a day if travelling east. This can lead to unusual adjustments. If you fly from Sydney, Australia on Saturday afternoon for about 14 hours, you arrive in Los Angeles in the morning, but *Saturday* morning as you have crossed the date line. So you seem to arrive before you leave!

If you travel significantly north or south of where you live you will find different stars, and the planisphere or star map you have at home will no longer be useful — you have to get another for that latitude. However, if you travel east or west without significantly going north or south, you can use your planisphere or star map, and the times marked on it will still be correct.

Kepler's laws

Let's imagine things from space again. The change in the Sun's diameter (see page 82) can be explained when we see that the Earth does not move around the Sun in a perfect circle,

---

Daylight saving time (summer time)

In order to make use of the early morning light after sunrise and use less artificial light by going to bed earlier in summer, thus saving energy, daylight saving time changes our clocks an hour forward in spring, and changes them back in autumn (hence the mnemonic 'spring forward, fall back'). It was introduced in Europe during the First World War, but was discontinued in many countries afterwards. It was used again in the Second World War and became widespread in the 1970s with the oil crisis. It is used in both northern and southern hemisphere countries (though obviously their summers are at opposite times of year), but is unnecessary in tropical countries, as there is not much difference between summer and winter daylight hours.

but moves in an *ellipse* (a kind of oval). In fact the ellipse is almost a circle, which is why it was only discovered relatively recently (hundreds of years ago, rather than thousands).

An ellipse has two centre points, called *foci* (one focus, two foci). If you firmly stick two pins a small distance apart, knot a string into a loop, and put the loop around the pins, you can draw an ellipse by putting a pencil inside the loop and drawing whilst keeping the string taught.

The Sun is at one of foci of the ellipse of the Earth's orbit. In January the Earth is closer, and in July further, from the Sun. Thus the Sun appears larger and smaller without actually changing size (Figure 60). Astronomers call the moment when the Earth is furthest from the Sun *aphelion* (from the Greek *apo*, away from, and *helios*, the Sun). And when the Earth is closest to the Sun it is at *perihelion* (from *peri*, close).

In 1609 Johannes Kepler (1571–1630), having analysed Tycho Brahe's (1546–1601) decades of accurate observations found that the planets (and the Earth) moved not in circles as

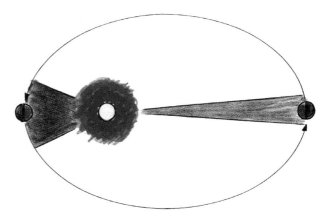

Figure 60. The elliptical orbit of the Earth around the Sun (greatly exaggerated). On the right is the July slow movement at aphelion, and to the left is January's fast movement at perihelion. The red areas are equal (Kepler's second law).

had previously been thought, but in ellipses. This is now called Kepler's first law.

He also discovered that when a planet (or the Earth) is at aphelion it moves more slowly, and when at perihelion it moves faster. He calculated that a line joining a planet and the Sun sweeps out equal areas during equal intervals of time (Figure 60). This is Kepler's second law.

Kepler's laws: Summary

- Kepler's first two laws states that:
  1. The orbit of every planet is an ellipse, with the Sun at one of the two foci.
  2. The line joining a planet and the Sun sweeps out equal areas in equal intervals of time.

- The consequence is that in July (at aphelion) the Sun appears to be a little smaller and moves a little slower; in January it appears a little larger and moves a little faster.

Figure 61. Rising Earth shadow with full Moon [Martin Rietze]

Earth's shadow

Looking east on a clear evening just after sunset, that is, with your back to the sunset, just above the horizon you can see a relatively sharp horizontal division in the sky with a dark blue below and lighter reddish colours above (Figure 61). The dark band gradually grows higher. The height of this division in the east shows how far the Sun is below the horizon in the west. Eventually the division disappears in the darkness of dusk. A similar thing can be seen before dawn in the west.

Just as any sunlit object casts a shadow, so the Earth too casts a shadow. But we can only see a shadow when it falls on something. The haze in the atmosphere is sufficient for us to see this shadow. The lower parts of the atmosphere are in darkness and so appear in dark blue shades, while the higher parts still catch the sunlight and appear lighter and reddish.

The shadow of the Earth can also fall on the Moon, causing an eclipse of the Moon; this will be described in greater detail later (page 123).

Sunspots

Photographs of the Sun show that there are spots on the surface, sometimes more, sometimes fewer, but rarely are there none. They appear singly or in groups and after a time disappear again (Figure 62). To see these, *never look directly at the Sun*. It is best to project an image of the Sun onto a sheet of paper, either through a pinhole or through binoculars. To view the Sun directly you must use a suitable filter — either welder's glass or aluminized Mylar. But these must *not* be used with binoculars or telescopes. (Lenses concentrate the invisible ultraviolet light that damages the retina, particularly in children's eyes.)

Roughly every eleven years there are a great number of sunspots. The last maximum was in 2012, and there may be another in 2023. Another thing that can be observed is that in the course of several days, the spots move across the surface of the Sun westwards (to right when viewed from the northern hemisphere). This shows that the body of the Sun rotates.

However, closer observation shows that spots around the equator of the Sun take about 27 days to go right around, while those closer to the Sun's poles take around 33 days. So the Sun is not a solid body like the Earth, but appears to be more fluid or gaseous.

Aurora

At high latitudes (60° and higher) like Alaska, northern Canada or Scandinavia, on some clear nights there is a wonderful display of coloured lights, appearing as wafting luminescent veils, always in silent motion. The colours can vary from white through yellow and green to red. First appearing gently, they can last for hours, and then disappear again. This is the *aurora borealis* or northern lights.

They are also visible in similar latitudes in the southern hemisphere, but there are no inhabited landmasses in such latitudes. (There they are called *aurora australis* or the southern lights.) Less often the aurora is visible at lower latitudes, occasionally even to 50° in Europe or 40° in North America, and in southern New Zealand and Tasmania and even Victoria in Australia. (The aurora are centred on the magnetic pole.) At times of sunspot maximum the aurora tends to be more frequent and widespread.

Rainbows

A much more common sight in temperate latitudes is a rainbow. It is not an astronomical phenomena, but it is intimately connected

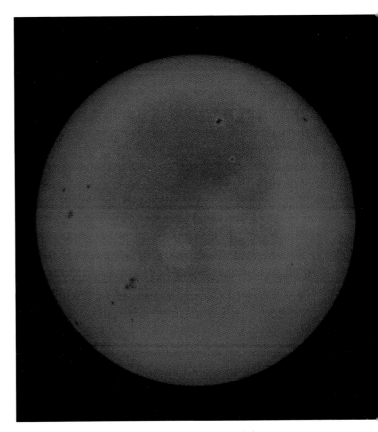

Figure 62. The Sun with many spots. [Wendelstein Observatory, Deutsches Museum, Munich]

with the Sun. It is visible opposite the Sun when it rains but the Sun shines. Why does it always appear as an arc? The colours can only appear at an angle of 42° from the point exactly opposite the Sun (called the *antisolar point*). As these points form a circle around the antisolar point, the rainbow forms a segment of a circle. If the Sun is rising or setting, the antisolar point is on or just below the horizon, and the arc of the rainbow will be very large (Figure 64). As the Sun gets higher, the antisolar point becomes lower, and the arc of

Figure 64. A rainbow at sunset. [Martin Rietze]

the rainbow becomes shallower. If the Sun is higher than 42° the antisolar point will be lower than 42° below the horizon, and the rainbow will not be visible.

The physics of a rainbow is as follows: light enters a raindrop and is reflected off the back of the drop. The light bends (refracts) before being reflected, and then bends again on exiting

Figure 63. The aurora showing veils of red and green light. [Cary Anderson]

the drop. The sunlight entering the raindrop contains all the colours of the spectrum, but each colour refracts a little differently on entering the more solid drop, and again on exiting it. Thus the sunlight is broken into the colours of the spectrum, with red appearing on the outside of the bow and purple on the inside.

Each one of us sees our own rainbow — if

we move to the left or right, the rainbow moves with us. If we try and reach it, we find it eludes us and is still some distance away, and if we go away from it, it follows us.

If you stay still and are lucky enough to be able to observe a rainbow for some time, you will find that the rainbow slowly moves to the right (in the northern hemisphere). This is because the Sun is slowly moving westwards, and thus the antisolar point (and with it the rainbow) moves eastwards. At sunset the effect is exaggerated if you look at where the right side of the arc touches the ground: because the Sun is setting, the rainbow is rising; this makes the arc bigger, moving the sides further apart. In the southern hemisphere where the Sun is to the north and the antisolar point to the south, the movement is to the left, and the left side exaggerates it. At dawn this is reversed.

PART III

THE MOON

Every little child has experienced that the Moon 'goes along with us' when we walk in moonlight. Houses and trees don't obstruct it, it faithfully follows our steps. And if there's a bigger thing in the way, we just keep going and it'll appear again in good time! Why does that happen? It is not only the Moon that does this — the stars and the Sun do the same. However, the stars are too small to notice, and the Sun is too bright. It's quite simple, really: the Moon is so far away, that we can't just walk past it; that's all there is to it.

On a stormy night when clouds scud across the sky, it seems as if the Moon is hurrying through the clouds; but that is merely an illusion — it is the clouds that are hurrying in front of the Moon.

People often wonder about is why the Moon is so much bigger when it rises, than later on when it's high in the sky. This too is an illusion: at the horizon we have trees, hills or houses to relate to the size of the Moon, but above us, these are missing. The Moon does very slightly change its size, but not during the course of a night. More of that later.

It is fascinating to look at the Moon through binoculars, particularly around the time of half Moon when the shadows highlight the mountains and craters.

The most obvious characteristic of the Moon is its changing shape — its phases. To understand the phases we shall first have to look at some other characteristics of the Moon.

Like the stars and the Sun, the Moon moves across the sky every day, rising in the east, culminating in the south (in the north in the southern hemisphere) and setting in the west. It moves against the background of the stars, remaining always in the constellations of the zodiac. Like the Sun during the year, the Moon moves through the zodiac in the opposite direction of the daily movement, but does this thirteen times faster than the Sun.

We can observe its movement very easily. We only need two successive clear nights when the Moon is near a bright star. On the first night look carefully where the Moon is in relation to the star, perhaps even making a quick sketch. On the following night the Moon will be further

Figure 65. The movement of the Moon in one day in relation to a star. The arrow shows the direction of this movement as seen in the northern hemisphere; in the southern hemisphere it moves in the opposite direction.

to the east by 15° — more than a hand's width (Figure 65). Taking the daily east to west movement of the Moon, together with its opposite movement through the zodiac, means that the Moon is slower in its daily movement than the stars. In fact it takes almost 25 hours to culminate again.

9. The Phases of the Moon

Northern hemisphere

The changing shape of the Moon is called its phases. We shall first describe these phases as seen in temperate northern latitudes. The new crescent Moon is first visible as a fine sickle in the west soon after sunset. The lit side is towards the Sun that has just set. We can see the crescent Moon for a short time before it also sets. In clear weather we can also see the ashen glow of the unlit part of the Moon; this is sometimes also called 'the old Moon in the new Moon's arms' (Figure 66). The sickle Moon has, of course, risen in the east, but as a fine sickle rising soon after sunrise, we can barely see it in the bright sky.

Every day the angle between the Moon and the Sun increases. At the same time, the

Figure 66. The new crescent Moon in the western evening sky. The ashen glow of the unlit part can often be seen as well also called 'the old Moon in the new Moon's arms' (northern hemisphere view).

crescent fills out and the time of the Moon setting becomes later. About five days after its first appearance the Moon appears as a half Moon (Figure 67). Astronomers call this the *first quarter*, because the angle between Sun and Moon is then 90°, a quarter of a circle. In this phase the Moon rises around noon and can be seen in the afternoon sky (the only reason we don't notice it, is that the sky is bright and we rarely look up). It culminates around sunset and sets around midnight. The lit side is still in the west (on the right).

Subsequently the Moon increases its angle from the Sun and continues to grow. This phase is called *gibbous* (from the Latin meaning hunch-backed). A week after first quarter it is *full Moon,* and it is opposite the Sun, rising at sunset (Figure 68), culminating at midnight, and setting at sunrise.

After full Moon, it continues in the same direction, moving east through the zodiac, but now the angle to the Sun reduces from the other side. The Moon shrinks in size, becoming gibbous again, but now the semicircular edge faces east. A week after full Moon it appears again half-lit, now rising at midnight, culminating around sunrise (Figure 69), and setting at noon. Astronomers call this the *third quarter* or *last quarter,* as the Moon is again 90° from the Sun.

In the following days the Moon continues its journey through the zodiac, reducing the angle to the Sun and appearing as a crescent that gets thinner every day, and rises later. About five or six days after the last quarter it appears for the last time just before sunrise as a fine sickle (Figure 70), before it disappears in the daylight.

Figure 67. The waxing half-lit Moon in the western evening sky, called first quarter (northern hemisphere view).

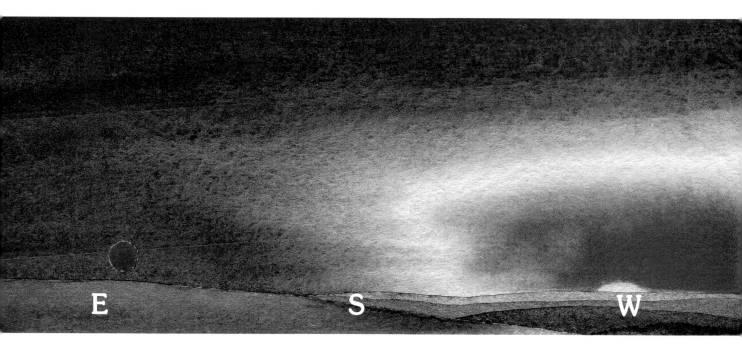

Figure 68. Full Moon rising at sunset.

Figure 69. The waning half-lit Moon in the eastern sky at dawn, called third quarter or last quarter (northern hemisphere view).

During the following three or four days the Moon remains invisible while it overtakes the Sun in the zodiac. Astronomers call the moment when Moon and Sun pass each other *new Moon.* So astronomers refer to something that cannot be seen at all. What in everyday language we call the new Moon — when the thin crescent first appears in the evening — is a day or two after astronomical new Moon, and is called a *new crescent Moon.* And then the whole cycle begins again.

In the two weeks between new Moon and full Moon, the Moon is said to be *waxing.* This has nothing to do with candle wax, but comes from an old Germanic word meaning 'to grow' (as in German *wachsen).* In the following two weeks the Moon is *waning.*

To help us remember whether the crescent of the Moon is waxing or waning, we can think of DOC. The curved side of the 'D' is on the left as the Moon grows, then we have full Moon, 'O', after which the left side is curved as the Moon is waning 'C'. Alternatively we can think of the curve of 'b' when the Moon *begins,* and 'd' when the Moon *declines.*

The Arctic and the North Pole

At the Arctic Circle the full Moon does not set in the long nights of midwinter. Roughly speaking, at midwinter when only half the Sun rises, only half the full Moon sets, for the full Moon is opposite the Sun. However, the Moon has 5° of 'freedom' so in fact the midwinter full Moon may be anywhere between 5° above or 5° below the horizon.

The Moon's daily movement at the North Pole is horizontal. As the Moon moves through

Figure 70. The waning crescent Moon in the eastern dawn sky (northern hemisphere view).

Figure 71. Full Moon at the North Pole in midwinter.

the zodiac, once a month it will be above the horizon for about two weeks, and then below the horizon for the following two weeks. In winter it will first become visible slowly rising as a waxing half Moon, reaching its highest as full Moon a week later (Figure 71); and then another week later, it sets as waning half Moon, and then disappears for two weeks.

In spring, the Moon rises as a thin waxing crescent, reaching its highest around waxing half Moon, and setting around full Moon. In summer, when the Sun is highest, the Moon rises as waning half Moon, is highest at new Moon (when we can see the crescent before and after new Moon), and then sets at waxing half Moon. And in autumn, around the annual sunset, it rises as full Moon and sets as the waning crescent.

At its highest the Moon is on average $23\frac{1}{2}°$ above the horizon, but because it can be 5° north or south of the ecliptic, the actual highest position of the Moon varies between $28\frac{1}{2}°$ and $18\frac{1}{2}°$.

The tropics and the equator
In the tropics the Moon is in the zenith once a month. However, because of its 5° tilt to the ecliptic it can sometimes be seen in the zenith as far north as latitude $28\frac{1}{2}°N$ (Orlando, FL, the Canary Isles, Delhi in India), and will always be in the zenith at some time during the month south of latitude $18\frac{1}{2}°N$ (Mexico City, Puerto Rico, Mumbai). In the southern tropics the southern limit of occasionally being in the zenith is at $28\frac{1}{2}°S$ (Easter Island, Florianópolis in Brazil, Kimberley in South Africa, Gold Coast in Queensland), and the limit when the Moon will

Figure 72. The setting new crescent Moon as seen in the tropics.

always be in the zenith at some time during the month runs through latitude of 18½°S (Harare in Zimbabwe, Townsville in Queensland).

At the equator the Moon also rises and sets vertically. As everywhere, full Moon is opposite the Sun, but the crescent Moon is noticeably different from other parts of the Earth: the waxing crescent lies like a bowl above the setting Sun (Figure 72), and similarly the waning crescent sits above the rising Sun. Because of the 5° tilt of the Moon's path to the ecliptic, it will not always be exactly above the rising or setting Sun, and so may be slightly inclined.

The southern hemisphere
In the southern hemisphere, the Moon waxes and wanes just like in the northern hemisphere. The waxing crescent is also visible in the evening, but is open to the right in the southern hemisphere, and the waning crescent of the early morning is open to the left.

To distinguish the waxing and waning crescent, remember COD. The curved side of the 'C' on the right is the waxing Moon, then full Moon 'O', after that the right side is curved as the Moon is waning 'D'.

The length of a month

The time it takes for the Moon to get back to the same star in the zodiac, called the *sidereal* period or month, is about 27⅓ days. This means that the daily movement of the Moon is about 13°. The time it takes for the Moon to come back to the same phase (full Moon to full Moon), called the *synodic* month, is 29½ days, and covers about 390° of the zodiac.

The overlap is owing to the Sun having moved 30° through the zodiac in the month it has taken for the Moon to return to the same star, and so the Moon has to continue for another 30° until it reaches the same angle with the Sun again. It is a bit like the hands on a clock. The larger (minute) hand returns to the same number after an hour, but then has to continue a little before it overtakes the smaller and slower hour hand. The ratio of speeds is roughly the same. In twelve hours, the minute hand passes the top twelve times, but overtakes the hour hand only eleven times. In a year the Moon passes a star thirteen times, but overtakes the Sun twelve times. (The clock hands, however, go in the opposite direction of the Sun and Moon's movement against the stars when viewed from the northern hemisphere.)

We saw earlier that the Moon's daily movement across the sky is slower than the Sun's. If the Moon rises at 8 pm today, it will rise at about 8:50 pm tomorrow. It can be a little earlier or later than that, but on average the Moon is late by about 50 minutes a day. (The Moon takes $27\frac{1}{3}$ days to go right round the zodiac, so every day it slips back by about $\frac{1}{27}$ of a day, which works out at about 50 minutes.)

We often come across paintings and drawings of a crescent Moon, where the artist has ignored elementary astronomy! A crescent Moon can only be visible after dusk or before dawn, and the lit part always points to the Sun. Figure 73 shows a charming New Year's card with the clock showing midnight, and with the Moon as a waning crescent but lit as if the Sun is high up in the sky.

Figure 73. An impossible New Year's scene: the fine crescent Moon cannot be above the horizon at midnight as the church clock shows, and the lit side points to a Sun high in the sky!

Another characteristic of the Moon is that the same side always faces the Earth. We might see the 'Man in the Moon' or something else — the important point is that the same side always faces the Earth. The Moon *rotates* around its axis at the same rate as it *revolves* around the Earth.

The Moon's Phases: Summary

- As well as the daily motion from east to west, the Moon moves through the zodiac from west to east. During this time its phases wax from crescent through half Moon and gibbous to full Moon, and then wane through gibbous, half Moon, crescent and then invisible as new Moon.
- The lit side of the Moon always points to the Sun.
- This synodic cycle takes about 29½ days.
- To return to the same star, the sidereal month, takes 27⅓ days.

- Moonlight increases and decreases in two ways: while the Moon is waxing, the length of time it is visible at night increases in the first part of the night. At full Moon it rises at sunset and sets at sunrise, shining all night long. While it wanes it is visible in the second part of the night, and the length of visibility decreases.
- Every day the Moon rises and sets about 50 minutes later.
- It always faces the same way towards the Earth.

Table of phases

Phase	Northern hemisphere	Southern hemisphere	Rises	Culminates	Sets
New Moon	●	●	sunrise	noon	sunset
Waxing crescent	◐	◑	mid morning	afternoon	early night
First quarter (half Moon)	◐	◑	noon	sunset	midnight
Waxing gibbous	◔	◕	afternoon	early night	late night
Full Moon	○	○	sunset	midnight	sunrise
Waning gibbous	◑	◐	early night	late night	mid morning
Last quarter (half Moon)	◑	◐	midnight	sunrise	noon
Waning crescent	◑	◐	late night	mid morning	afternoon
New Moon	●	●	sunrise	noon	sunset

A lunarium

You can make a little model, a lunarium, to help understand the phases. Apart from a little patience you only need some card (preferably blue and a dark colour), some white and yellow paper, a split pin and some adhesive. For tools a pair of compasses, scissors and fine felt-tipped pens (black and ideally white).

With the compasses draw a circle of 7.5 cm (3 in) radius on the blue card and cut it out. That is the sky. Cut the horizon out of the dark card as shown in Figure 74. Cut a disc about 10 mm (½ in) diameter out of white paper — this is the Sun. Cut 5 similar sized discs out of the yellow paper — these are the Moon.

Now with the black pen blacken parts of the yellow discs to make two crescent and two half Moons. The full Moon does not need any blackening. Stick the Sun and Moon discs in the right positions as shown in Figure 75. Make sure the full Moon is exactly opposite the Sun

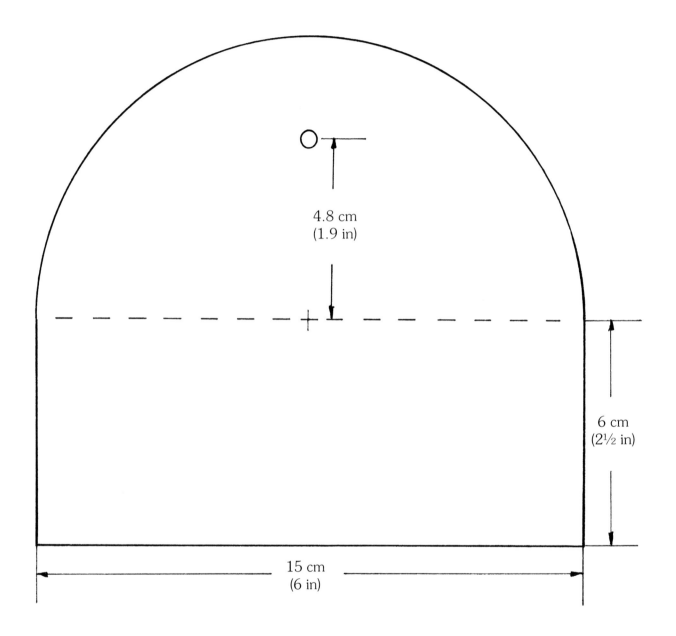

4.8 cm
(1.9 in)

6 cm
(2½ in)

15 cm
(6 in)

Figure 74. Cut out the horizon card in dark card.

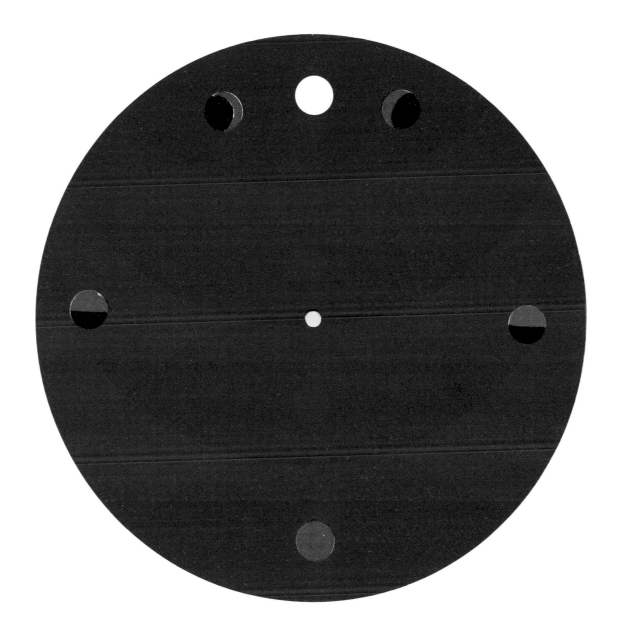

Figure 75. Cut out the sky card in blue card, 15 cm (6 in) diameter.

Figure 76. The finished lunarium.

and the half Moons exactly at right angles. If you want to, you can add a two gibbous Moons between full Moon and the half Moons, making sure the darkened part is away from the Sun. Pierce a hole in the centre of the blue disk and in the marked place of the horizon disc.

Figure 76 shows the finished lunarium with east, south and west marked for the northern hemisphere. Turn the disc clockwise to show the daily movement. This lets you see when and where the Moon rises in relation to the Sun at different phases. If the white Sun is visible, it is daytime. This lunarium shows five different phases (or seven if you've added gibbous Moons) at once. Of course in reality there is only one Moon and one phase at a time, but it saves making a lunarium for each phase!

For a southern hemisphere lunarium mark west, north and east (instead of E S W) and turn the disk anticlockwise to show the daily movement.

Armillary sphere

It can be difficult to visualise the complex movements of the stars, Sun and Moon in different parts of the world. A three-dimensional model, an *armillary sphere,* can help. Such an instrument was used by navigators and astronomers of the past and can be found in museums. It has a movable celestial equator and ecliptic and a fixed horizon, and can be set for any latitude (Figure 77). The inner, movable part shows the geocentric movement of the zodiac. The Sun, Moon and planets can be marked on the zodiac.

Figure 77. An armillary sphere [AstroMedia]

The Moon viewed from space

The phases can easily be understood when imagined from space. The Moon revolves around the Earth; it is sometimes called a *satellite* of the Earth. The direction of orbit is the same as that of the Earth around the Sun, but it is much closer to the Earth (Figure 78). The Moon has no light of its own: it simply reflects sunlight. Half the Moon is lit by the Sun. In position 1 from Earth we are looking at the unlit side of the Moon: it is new Moon. Position 2 is the first quarter or waxing half Moon. Seen from the northern part of the Earth, the right side of the Moon is lit. In position 3 the Moon is opposite the Sun — full Moon — and we

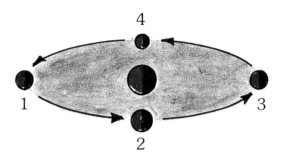

Figure 78. The Moon's phases are caused by its orbit around the Earth, and our seeing different proportions of its sunlit side. 1 new Moon, 2 first quarter, 3 full Moon, 4 last quarter.

see the entire lit side. Position 4 is the waning half Moon, or last quarter. The idea that the Earth's shadow causes the Moon's phases is a misconception.

Does the Moon also rotate? The fact that we always see the same side of the Moon from Earth shows it has a rotation of exactly the same length as its revolution (29½ days in relation to the Sun).

A little girl once asked what would happen if you built a house on the Moon, and it then waned. Of course, the Moon does not shrink; we simply see less of the lit side every day while it wanes. The house would thus simply be in darkness, and its fictional inhabitants would have seen the Sun set.

The average length of a calendar month is 30½ days, one day longer than the synodic rhythm of the Moon. This means that the date of a particular phase is earlier by one day every month. So for instance if it is full Moon on July 12, the next fill Moons are on August 10, September 9, October 8, November 6 and December 6. While the average is one day earlier, we can see in fact the date jumps around a little, for the calendar months vary in length, there are leap years and, as we shall see, the Moon itself moves a little faster or slower.

Twelve Moon months, or a Moon 'year', are thus about twelve days shorter than a calendar year which is based on the Sun. Generally every year has twelve full Moons and twelve new Moons, though about every third year has thirteen.

The Moon's Phases Seen from Space: Summary

- The Moon is like a freely suspended sphere lit by the Sun. The lit side always faces the Sun.
- The Moon orbits the Earth while the Earth orbits the Sun. The Moon's orbit is much smaller than the Sun's. While the Earth takes a year to go around the Sun, the Moon just takes a month to go round the Earth.
- The Moon's orbit is roughly in the same plane as the Earth's orbit around the Sun, so we always see the Moon in a constellation of the zodiac.
- The Moon rotates around an axis perpendicular to its plane of orbit, at the same speed as its revolution, so that we always see the same side from the Earth.
- The Moon's waxing or waning is caused an increasing or decreasing part of the lit side facing the Earth.

10. The Moon in the Zodiac

To recap, we saw the Moon is always in one of the constellations of the zodiac. As well as its daily movement from east to west like the stars, there is the opposite movement against the background of the stars. This movement is much faster than the Sun's. The Moon takes $27\frac{1}{3}$ days while the Sun takes a whole year. During the year the Moon moves through the zodiac about 13 times.

Put another way, as well as their daily movement, the Sun and Moon pass through the zodiac. The Moon moves much faster, so that the Moon keeps overtaking the Sun, twelve times in a year. Every time the Moon overtakes the Sun they are close together: it is new Moon. Exactly between these meetings they are opposite each other, and it is full Moon.

While on its path through the zodiac the Moon meets many stars. Of course it does not push them aside, but passes in front of them, blocking their light. Astronomers call this an *occultation*. Occultations of brighter stars or even planets occur sometimes — Aldebaran, Regulus, Spica and Antares may be hidden by the Moon. Occasionally there is an occultation of a planet by the Moon. Even the Sun can be occluded by the Moon: that is an eclipse, and

we shall look at these later (page 119). But let us return to the Moon in the zodiac.

The northern hemisphere
In its wanderings through the zodiac the Moon, just like the Sun, is above the celestial equator for half the time, and below it for the other half. As it takes about 4 weeks to go through the zodiac, for about 2 weeks it is in the northern part of the zodiac and for 2 weeks in the southern part. It does not suddenly change, but, as with the Sun, the change is gradual. From when the Moon is in the lowest part of the zodiac (between Scorpius and Sagittarius) it is *ascending* until it reaches the highest part (between Taurus and Gemini). For the other half of the month it is *descending.* This has nothing to do with the Moon's altitude (height above the horizon) between its daily rising and setting.

Ascending and descending in the zodiac each takes about 2 weeks. This is quite independent of the Moon's phase. In winter, in December, the waxing phase is simultaneously the time of ascending Moon, while in June it is the opposite: the waxing phase is during the time of descending Moon. In between times there is a transition.

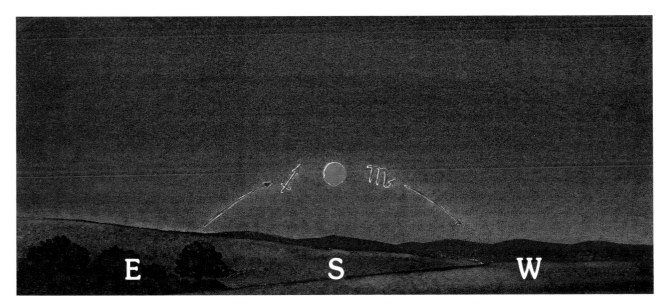

Figure 79. The low arc of the summer full Moon between Scorpius and Sagittarius (northern hemisphere).

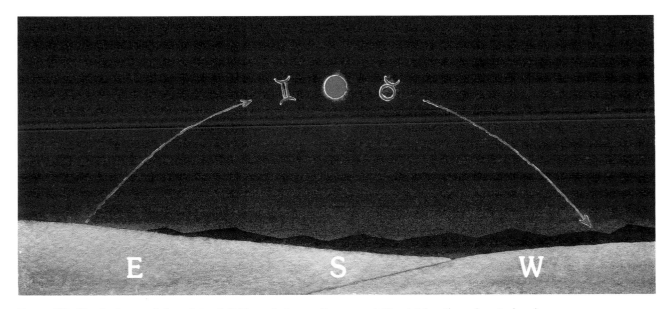

Figure 80. The high arc of the winter full Moon between Taurus and Gemini (northern hemisphere).

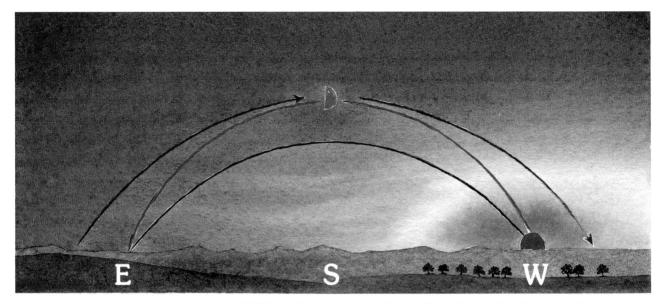

Figure 81. A spring evening with a waxing half Moon (northern hemisphere). The zodiac (ecliptic) is in a high position (yellow) above the celestial equator (red). The Moon's daily arc is at its greatest (purple).

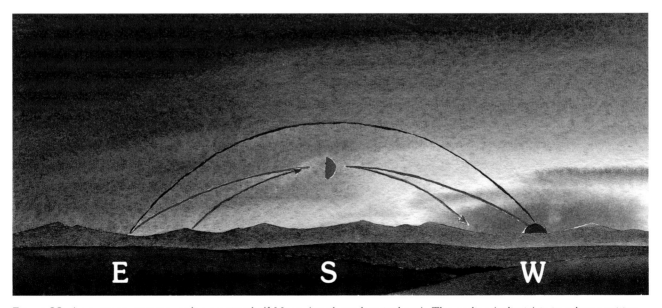

Figure 82. An autumn evening with a waxing half Moon (northern hemisphere). The zodiac (ecliptic) is in a low position (yellow) below the celestial equator (red). The Moon's daily arc is at its smallest (purple).

Figure 83. The new crescent Moon is higher and more horizontal in spring, and lower, more vertical in autumn (northern hemisphere).

In its daily course, the Moon moves in sometimes greater and sometimes smaller arcs through the sky. If it is on a high arc, it rises towards the north of east and sets towards the north of west, and is visible for a long time in the sky; while on a low arc, it rises and sets closer to the south of east and west. With the Sun, this difference in daily arcs determines the seasons, but with the Moon it has no effect on the seasons, as the Moon goes through this cycle twelve times a year.

Full Moon is opposite the Sun. In June the Sun is in the highest part of the zodiac, so the full Moon is in the lowest part of the zodiac, between Scorpius and Sagittarius (Figure 79). During the short summer nights the full Moon has a low arc across the night sky.

Figure 80 shows the full Moon at the highest part of the zodiac, between Taurus and Gemini, in the long December nights rising well to the north of east and setting well to the north of west, and taking well over twelve hours to cross the sky (the exact time depends on our latitude). It is worth observing exactly where the full Moon rises and sets at these extreme times, for it reminds us where the Sun rises and sets six months earlier or later.

In spring evenings the zodiac is in a high position. At this time of year it is the waxing half Moon which is high in the sky (Figure 81). In autumn evenings the zodiac is in a low position, and the waxing half Moon is low in the sky (Figure 82).

The waxing crescent Moon is sometimes more vertical, sometimes more horizontal (Figure 83). Again it is a matter of the position of the zodiac. When it is high, the Moon will be higher and incline more to the horizontal. This is the case in spring. In autumn evenings the zodiac is in a low position, so the Moon

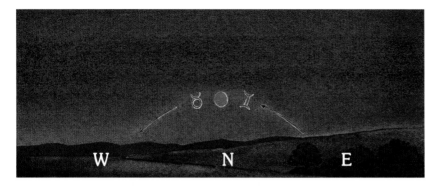

Figure 84. The low arc of the summer (December) full Moon between Taurus and Gemini (southern hemisphere).

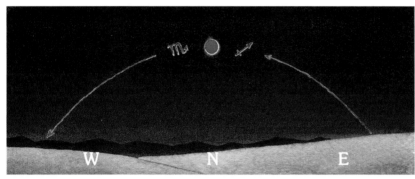

Figure 85. The high arc of the winter (June) full Moon between Scorpius and Sagittarius (southern hemisphere).

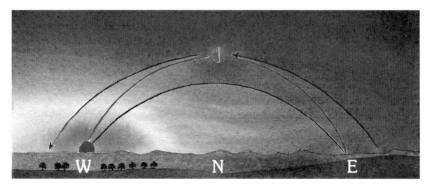

Figure 86. A spring evening (September) with a waxing half Moon (southern hemisphere). The zodiac (ecliptic) is in a high position (yellow) above the celestial equator (red). The Moon's daily arc is at its greatest (purple).

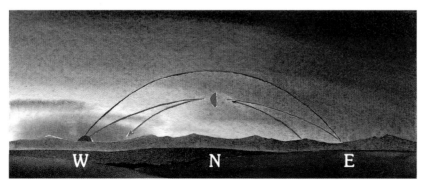

Figure 87. An autumn evening (March) with a waxing half Moon (southern hemisphere). The zodiac (ecliptic) is in a low position (yellow) below the celestial equator (red). The Moon's daily arc is at its smallest (purple).

Figure 88. The new crescent Moon is higher and more horizontal in spring, and lower, more vertical in autumn (southern hemisphere).

will be lower and more vertical. At dawn the waning crescent appears similarly, but is high in autumn and low in spring.

The southern hemisphere

Here too the Moon is above the celestial equator for about 2 weeks, and below it for another 2 weeks. It is in the lowest (northern) part of the zodiac between Taurus and Gemini, and is *ascending* until it reaches the highest part between Scorpius and Sagittarius. It then becomes *descending* in the opposite part of the zodiac. So in the southern hemisphere the descending and ascending Moon is always the opposite of the northern hemisphere's Moon.

Ascending and descending in the zodiac each takes about 2 weeks, independent of the Moon's phase. In winter, in June, the waxing phase coincides with the time of ascending Moon, while in December the waxing phase is during the descending Moon. (The phases are the same in both hemispheres, but ascending and descending are opposite).

The full Moon is opposite the Sun, so in December the Sun is in the highest part of the zodiac and the Moon is in the lowest part, between Taurus and Gemini. During the short summer nights the full Moon has a low arc across the night sky (Figure 84).

When the full Moon is at the highest part of the zodiac (Figure 85), between Scorpius and Sagittarius, in the longer June nights it rises to the south of east and sets to the south of west, taking well over twelve hours to cross the sky (the exact time depends on the latitude).

In spring evenings (September) the zodiac is in a high position (Figure 86). At this time of year the waxing half Moon is high in the sky. In autumn evenings (March) the zodiac is in a low position, and the waxing half Moon is low (Figure 87).

In the southern hemisphere the waxing crescent Moon in the evenings will be high and more horizontal in spring (September) when the zodiac appears in its high or steep position. In autumn (March) it is low and more vertical. At dawn the waning crescent appears high in autumn and low in spring (Figure 88). This is the same as in the northern hemisphere because the seasons are opposite.

The Moon in the Zodiac: Summary

- As well as its daily movement, the Moon ascends and descends through the zodiac. This ascending and descending is independent of the phases.
- The summer full Moon moves in a low arc in the short summer nights, similar to the Sun's winter arc.
- The winter full Moon has a high arc like the summer Sun.

---

11. Nodes and Eclipses

The lunar nodes

By now we should be wondering why, every time it is new Moon, the Moon does not block out the Sun's light causing an eclipse. The reason is that the Moon does not move through the zodiac exactly on the ecliptic, the Sun's path. Its path is inclined to the ecliptic by 5°, half above and half below the ecliptic. So when the Moon passes the Sun at new Moon, it is usually a bit above or below the Sun.

However, during the course of a month, the Moon will cross the Sun's path twice, once descending from above and once ascending from below the ecliptic. This happens independently of the Moon's phase or its position in the zodiac. The Moon's orbit being inclined to the Sun's path (ecliptic) by 5° is similar to the ecliptic being inclined to the celestial equator by 23½°. You could say that the Moon's orbit has a double inclination — 23½° to the equator (like the Sun) and additionally 5° to the Sun's path, the ecliptic.

Let us look at it from another viewpoint. Like the Sun, the Moon moves eastwards through the zodiac, but much faster. It has its own path which is up to 5° away from the ecliptic. For half a month the Moon is above (in the northern hemisphere to the north of) the ecliptic, and for the other half it is below (to the south). Therefore every two weeks its path crosses the ecliptic, once from north to south, once from south to north. Astronomers call these crossing points of the Moon's path and the ecliptic *nodes*. There is an *ascending node* when the Moon crosses into the higher, northern part of the zodiac, and a *descending node* when the Moon crosses into the lower, southern half of the zodiac. (Even though seen from the southern hemisphere the nodes are the other way round, astronomers *always* mean the point when the Moon crosses into the northern part

---

Signs for the nodes

Sometimes the following signs are used for the nodes:
☊ is the ascending node.
☋ is the descending node.

Figure 89. An example of the Moon's path (red) through the zodiac (northern hemisphere view). In Cancer (♋) and Leo (♌) the Moon is below the ecliptic (yellow); it crosses the ecliptic in Virgo at the ascending node, and is above the ecliptic in Libra (♎), Scorpius (♏) and Sagittarius (♐).

of the zodiac when using ascending node. Some call this the north node, and the descending node the south node, to avoid any confusion.) The two nodes are opposite each other.

So usually the new Moon passes the Sun to the north or south. Sometimes, however, it passes in front of the Sun. This happens if the new Moon is just crossing the ecliptic, that is, if it is just at a node. Regardless of whether the node is an ascending or a descending one, there will be an eclipse. The times of a possible eclipse are half a year apart as the nodes are opposite each other. Figure 89 shows an example of the Moon at a node. Its path (red) through the zodiac shows it coming from below the ecliptic (yellow), crossing the ecliptic at the

ascending node, and then continuing above the ecliptic in the following constellations of the zodiac.

The movement of the nodes

There is yet more to understand about the nodes, because we have not said where in the zodiac the nodes are. This is for the simple reason that the nodes move. They move in the opposite direction to the Moon's movement through the zodiac. They take 18.6 years to move right round the zodiac. Figure 89 shows when the ascending node happened to be in Virgo.

As the nodes move in the zodiac, so does the position of the Moon's highest angle (5°)

above the ecliptic, and opposite that, the lowest position. If the ascending node coincides with the spring equinox point, at the highest part of the ecliptic, the Moon's path will be even higher by 5°, and the lowest point of the Moon's path will be 5° lower than the lowest point of the ecliptic. (The result in the southern hemisphere will also be extreme: when in the northern hemisphere the Moon is between Taurus and Gemini, at the highest part of the zodiac, it is 5° higher than — further north of — the ecliptic, in the southern hemisphere the same moment is when the Moon is in the lowest part of the ecliptic and it will be 5° lower than — further north of — the ecliptic.)

If, however, the ascending node coincides with the autumn equinox (and the descending node at the spring equinox) the Moon's path will be less steep than the ecliptic. The highest position of the Moon will be 5° less than the position of the summer Sun, and the lowest position will be 5° higher than the Sun's lowest. From one extreme to another takes 9.3 years.

What does this look like in practice? We may notice it around the time of full Moon. Initially we said that a winter midnight full Moon appeared as high as the summer noon Sun. However, we have now seen that this could be 5° higher or lower. Similarly we can see the same with the summer full Moon, where its low arc may be even lower, or may be higher.

Unless we look out for these subtle differences in the Moon's path we will not notice them. In 2006 the ascending node was at the spring equinox, so the Moon's highest and lowest positions were extremely high and low. This will be repeated in 2025. Halfway between these two years, in 2015, the ascending node is at the autumn equinox, and so the Moon's highest and lowest positions will be least high and low. This moderate pattern will happen again in 2034.

Solar eclipses

When new Moon is at a node, the Moon will cover the Sun. A total eclipse can only be seen from a small part of the Earth, and at best up to 7½ minutes. During this time, a mysterious light is visible around the Sun, the *corona,* and at the edges some reddish *prominences* (or *protuberances),* bright flame-like forms may be seen (Figure 90). The sky is dark, the surroundings are grey, and the air cools. The stars appear. It is a very moving experience and the only opportunity to see the stars of the constellation in which the Sun is at that moment. During the few minutes of totality you do not need protective glasses, but in the time before and after — even if only a tiny bit of the Sun is still visible — you must use a suitable filter — either welder's glass or aluminized Mylar.

A total eclipse cannot be seen from everywhere on Earth. The zone of totality is narrow (at most 260 km, 160 miles wide) and sweeps across the dayside of the Earth. To the sides of this zone the eclipse will only be partial; that is, the Sun is only partly covered rather like before and after a total eclipse. A partial eclipse can last several hours, though it hardly gets darker, and the corona and stars will not be visible.

Figure 90. The totally eclipsed Sun with corona and protuberances. Eclipse of August 11, 1999, from southern Germany. [Martin Rietze]

Figure 91. A partial solar eclipse. This series of photographs was taken at about 5-minute intervals. The Sun and Moon are setting towards the west (right), while the Moon is moving eastwards in relation to the stars. [Martin Rietze]

Figure 92. An almost total lunar eclipse. [Martin Rietze]

Figure 91 is a series of photographs showing the progress of a partial eclipse. The Sun and Moon are setting towards the right (west). It looks as if the Sun is overtaking the Moon, but against the background of the stars, the Moon is moving eastward much faster than the Sun.

If you have the opportunity to observe a partial eclipse (with suitable glasses — see above), it is worth looking (without the glasses) at the shadow pattern on the ground. If you are under a leafy tree you will see lots of little moon-shaped lights between the shadows of the leaves. This is a pinhole-camera effect. You can produce such a shadow by making a pinhole in a card. Instead of a little round light patch, it will be moon-shaped during an eclipse (but upside-down and back-to-front).

Incidentally the shape of the Sun during an eclipse (or its projected pin-hole image) is different from that of a crescent Moon: the inside line of a crescent Moon is in the shape of a semi-ellipse, while that of a partial eclipse is a semicircle, the edge of the Moon. We can see this looking at the shapes of the Sun in Figure 91, which are unlike the crescent Moon.

Lunar eclipses

Occasionally during a full Moon night we can experience another drama: a lunar eclipse. The Moon enters the otherwise invisible shadow of the Earth, because the full Moon is passing a node. In contrast to a solar eclipse, a lunar eclipse is seen from all over the nightside of the Earth — because the full Moon is exactly opposite the Sun, it is above the horizon (and thus visible) wherever the Sun is below the horizon. At the distance of the Moon, the shadow of the Earth is about 2½ times as big as the Moon, so the time of totality lasts up to 100 minutes if the Moon passes through the centre of the shadow. During totality the Moon appears in as an orange-red or even dark brown-grey disc (Figure 92).

The Moon moves from west to east through the earth's shadow, beginning as a partial eclipse before becoming total, and then ending again as a partial eclipse. From the beginning of the partial to the end of the closing partial phase of the eclipse can take up to 3 hours and 40 minutes. The shadow border (called *terminator* by astronomers) is not as sharp as might be expected, due to some sunlight reaching the Moon through the Earth's atmosphere.

Lunar eclipses occur two weeks before or after a solar eclipse, though there is not always a lunar eclipse before or after a solar one. The lunar eclipses may also be partial.

Nodes and Eclipses: Summary

- The Moon does not move through the zodiac exactly on the ecliptic, but diverges up to 5° from it. Its orbit is inclined to the ecliptic by this amount, and crosses the ecliptic in two opposite nodes.
- The nodes are independent of the Moon's phase, and moves through the zodiac in 18.6 years in the opposite direction of the Moon's movement in the zodiac.
- If new Moon occurs close to a node, a solar eclipse results.
- If full Moon is close to a node, a lunar eclipse results.

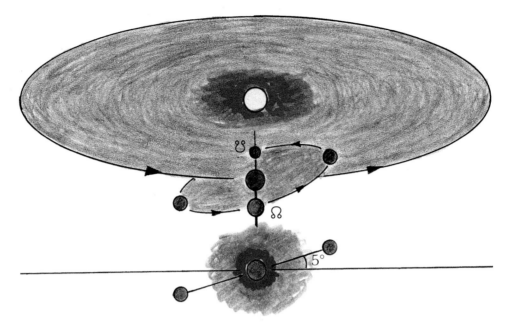

Figure 93. The plane of the lunar orbit (purple) is inclined at 5° to the Earth's orbital plane (brown). The nodal line connecting descending and ascending nodes (☋ and ☊) points to the Sun, indicating that there will be a solar eclipse at new Moon and lunar eclipse at full Moon. Below: end-on view of the Sun with the Earth and the Moon shown from their night side.

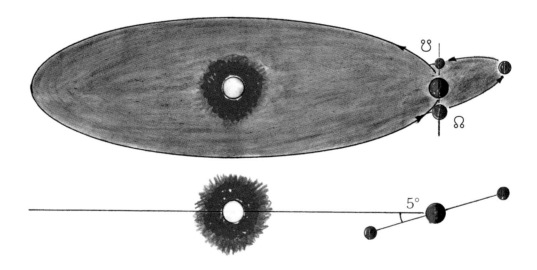

Figure 94. The plane of the lunar orbit (purple) is inclined at 5° to the Earth's orbital plane (brown). The nodal line connecting descending and ascending nodes (☋ and ☊) does not pass through the Sun, indicating that there will not be eclipses. Below: end-on view of the Sun illuminating the Earth and the Moon.

The lunar nodes viewed from space

The changing position of the Moon above and below the ecliptic can easily be understood when imagined from space.

If the orbit of the Moon was in the same plane as the orbit of the Earth (the ecliptic), there would be a solar eclipse at every new Moon. But that is not the case, for the lunar orbit is inclined at 5° to the ecliptic. The two orbital planes intersect in the *nodal line* that connects the ascending and descending nodes. Figure 93 shows the orientation of the nodal line when eclipses occur. Figure 94 shows the orientation of the nodal line three months later, with the full Moon shown above the ecliptic and the new Moon below. Another three months later the nodal line is again orientated towards the Sun and eclipses will occur. So the times of eclipses are six months apart.

With a solar eclipse the shadow of the Moon falls on the Earth. With a lunar eclipse the shadow of the Earth falls in the Moon.

The nodal line does not remain stationary in the zodiac, but revolves around the Earth from west to east through the zodiac in 18.6 years. That causes the times of eclipses to be about 10 days earlier each year, so they are not tied to certain months or seasons.

12. The Apparent Size of the Moon

The Sun and Moon appear in the sky to be about the same size. That, of course, is the reason why a total eclipse is so short. However, there are some eclipses which do not appear to be total, as the Moon is too small to entirely cover the Sun. A ring of the Sun appears all around the darkened part. This is called an *annular eclipse* (from Latin *annulus*, a ring). At other times the Moon appears larger and the time of totality can last for 7½ minutes. So the size of the Moon's disc changes, though this does not happen very quickly.

With the unaided eye this difference in size is only seen during an eclipse when the Moon is directly in front of the Sun. This of course does not happen very often, and to follow all the solar eclipses we would have to travel all over the world to be in the right place each time.

The actual distance to the Moon can be measured by observing the angle of the Moon from two or more distant places on the Earth with accurate telescopes, as the angle will be slightly different to this relatively close celestial body. The distance to the Moon is about 30 times the diameter of the Earth, about 385 000 km (239 000 miles), a distance which would take a healthy person a lifetime to walk. We can

Rhythms of the Moon

The **synodic** month is the time taken from one phase back to the same one. On average it is about 29½ days (29^d 12^h 44^m).

The **sidereal** month is the time taken to return to the same star. On average it is about 27⅓ days (27^d 7^h 43^m).

The **nodal** month is the time taken to return to the same node (crossing point with the ecliptic). On average it is about 27¼ days (27^d 5^h 6^m).

The **apsidal** month is the time taken to return to the same position in the elliptical orbit (for instance, apogee to apogee). On average it is about 27½ days (27^d 13^h 19^m).

then calculate the Moon's actual size can which is found to be about a quarter of the Earth's.

As we have said, the size and distance of Sun and Moon are arranged in such a way that they appear to be the same when seen from the Earth. Whether a coincidence or by design, it is something we can marvel at. As the Sun is 400 times further than the Moon, it must about 400 times larger, or 100 times larger than the Earth.

The changes in the diameter of the Moon seen from the Earth can be explained by the fact that the Moon's orbit around the Earth is not a circle, but an ellipse. Kepler's laws hold for the Moon as well: the Moon moves in an ellipse, and the Earth is at one of the foci of the ellipse. Astronomers call the moment when the Moon is closest to the Earth *perigee*, and the opposite point, when it is furthest away, *apogee*. From one position to the opposite takes about two weeks. When the Moon is at apogee it moves a little slower, and at perigee a little faster. The line connecting perigree and apogee is called the aspidal line.

Perigee and apogee do not remain in the same place in the zodiac, but also slowly move eastwards, as the time from one perigee to the next is a little longer than the sidereal month (returning to the same star).

A total solar eclipse lasting for a long time occurs when the Sun is at aphelion (in July) and the Moon at perigee. When the opposite happens (Moon are apogee and Sun at perihelion) there is an annular eclipse with a large, visible ring.

Summary

- The apparent size of the Moon varies during its monthly course through the zodiac by about 14% (one seventh) of its diameter.
- This means that the Moon's orbit around the Earth is an ellipse. This ellipse a little is flatter than the ellipse of the Earth's orbit around the Sun.
- The variations of the Moon's apparent size are about 4 times as much as the Sun's.

13. Other Aspects of the Moon

The tides

The Moon has a great influence on the oceans. Anyone living near the sea will be familiar with the tides. The Moon attracts the water under it, causing a kind of swelling of the ocean which we see as high tide. As the Earth rotates, so the high tide rotates around the Earth, and as the Moon is also moving, high tide arrives a little later every day.

The Moon attracts the ocean's water closest to it, and leaves the water on the opposite side

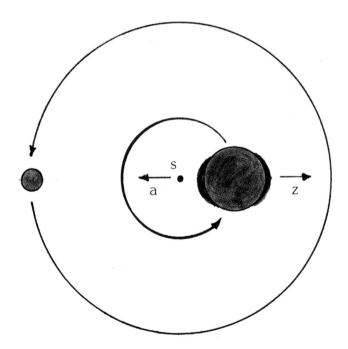

Figure 95. The Moon attracts the water closest to it a little, and leaves the water on the opposite side 'behind', because the gravitational pull is weaker further away.

of the Earth 'behind' as it were, because the gravitational pull is weaker when further away (Figure 95).

While this looks very simple, in detail it is very complex. Most Atlantic Ocean coasts have two high tides and two low tides every lunar 'day' of 24 hours 50 minutes. Some Pacific Ocean coasts have only one high tide and one low tide every lunar 'day'. Many places have a combination of these, and so have uneven tides (for instance a very high tide, a less low tide, a less high tide, followed by a very low tide). The tides are less pronounced in the middle of an ocean (for instance, on islands in the Pacific) than on the far coasts of the ocean basin.

The Moon is also said to influence plant growth, animals and even human behaviour, but to go into that would go far beyond the scope of this book.

The date of Easter

The Sun and Moon have an important role in fixing the date of Easter. Three conditions have to be met for Easter, in the following order. First the spring equinox has to pass, so the Sun is in the higher part of the zodiac. Secondly, full Moon has to pass; that full Moon is called the Easter (or Paschal) full Moon. Thirdly, the right day of the week has to come: Sunday. So the simple rule is the first Sunday *after* the first full Moon *after* the spring equinox.

Easter Sunday therefore always falls between March 22 (if the Easter full Moon is on the day of the equinox, and that day is Saturday) and April 25 (if the Easter full Moon is on a Monday 29 days after the equinox, on April 19, the following Sunday will be April 25).

Contrasts of the Sun and Moon

We have seen how complicated the movement of the Moon is, and it is in fact even more complex than we have described. It is difficult to accurately predict the Moon's position as it is so subtle and sensitive. In contrast the movement of the Sun is very simple.

The surface of the Moon, even under the most exacting astronomical observations, shows practically no change. It is a barren lifeless desert. In contrast, observation of the Sun's surface shows a continually changing pattern — spots appear and disappear, huge flares light up for a few hours and die away. It is unpredictable and full of life.

Our two great heavenly bodies are very different.

PART IV

PLANETS, COMETS AND METEORS

14. The Planets

Among the 2500 stars that one can see with the unaided eye on a clear night, there are five that change their positions in relation to the other stars. They remain within the zodiac, and have a steady light (particularly just above the horizon where other stars twinkle). These are the planets. It is seldom that all five can be seen at once; even to be able to see three or four together does not happen very often.

Because the planets move among the 'fixed' stars, they cannot be marked permanently on a star map. If you see a bright star in the zodiac that is out of place when compared to a star map (or what you remember from some months ago), you can be fairly certain it is a planet. Their movement is slow and steady — from one night to the next you will not be able to see any difference, but it becomes clear after several weeks or months. They usually move eastwards in the zodiac (in the same direction as Sun and Moon), but their speed varies. In fact they sometimes become stationary and move 'backwards' (westwards)

through the zodiac for a time. Astronomers call this movement *retrograde,* in contrast to the usual *direct* movement.

Like the Moon, the planets do not move exactly on the ecliptic but slightly above or below it, and they have nodes where their orbit crosses it. During their retrograde movement they make a loop against the stars and change their brightness.

Each planet has a characteristic movement. Mars, Jupiter and Saturn make a loop each time when they are in *opposition* (that, is opposite the Sun in the zodiac). Venus and Mercury also make loops, but are never in opposition. These two are always close to the Sun, sometimes ahead and sometimes behind. When the Sun and the planet pass each other they are in *conjunction*. This is similar to new Moon, and the planet is not visible at this time. The first group (Mars, Jupiter, Saturn) are called *superior* planets, while Venus and Mercury are *inferior* planets. These terms are used in their original meaning of 'greater than' and 'less than' (the orbit of the earth), and not in today's common 'better' and 'worse' sense.

There are further planets which can only be seen with the aid of a telescope: Uranus (discovered in 1781) and Neptune (discovered in 1846). Beyond that are a number of what are now called dwarf planets, the best known being Pluto (discovered in 1930). As this book describes what we can see with the unaided eye, we shall not say more about these planets.

In the following descriptions, we shall ignore the daily motion, and look at the movement of the planets in relation to the stars and the Sun.

The planets are all visible in the southern hemisphere, but like Sun and Moon move from left to right through the zodiac (except while they are retrograde). This is still an eastward movement as described earlier, but when facing north the movement appears the other way round.

Mars

The ancient Greeks and Romans saw Mars as the god of war. Mars has a reddish light and makes varied loops. We shall look at some examples. Mars moves at roughly half the speed of the Sun through the zodiac, so takes about two years to complete its orbit. However, sometimes it slows down, becomes stationary and then moves retrograde (westwards). In the middle of the retrograde motion it is exactly opposite the Sun and becomes very bright.

Figures 96–99. Various loops of Mars. In the middle of the retrograde motion Mars is at its brightest. The thickness of the lines give an impression of the movement in three dimensions.

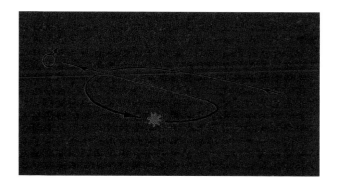

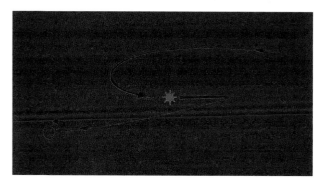

This is when it is visible all night long. The retrograde motion is not on the same path as the direct motion, so it creates a loop. This loop can be downwards or upwards (Figures 96, 97), or even a S- or Z-shape (Figures 98, 99). Mars never makes the same shaped loop twice, and rarely makes a symmetrical one.

On average the distance of retrograde motion is about 15°, half a constellation. The time of retrograde motion is two to three months. If you include the direct motion over that distance before and after the retrograde, it takes almost six months. That is the time Mars is most visible, and it repeats this pattern every two years and two months, though it is not very regular.

After completing the retrograde motion, Mars loses brilliance and is only visible in the evenings. The Sun, moving steadily through the zodiac, is beginning to catch up with Mars. Having lost so much time during the loop, Mars now rushes forward ever faster, but not fast enough to avoid the Sun. As it gets closer to the Sun, Mars disappears in the evening twilight.

Very slowly the Sun passes Mars: the *conjunction*. We are unable to see this as Mars rises in the morning and sets in the evening with the Sun. Eventually, after six months, Mars appears again in the dawn twilight. Gradually Mars becomes brighter as the distance to the Sun grows. Its movement in relation to the stars begins to slow down as if exhausted from the fruitless running from the Sun. And then the whole cycle of events begins anew.

To see such a loop it is best to make a careful sketch of the stars around Mars, and note the date of the sketch. Making a sketch at weekly or two-weekly intervals will ensure you begin to see the pattern. Check in an almanac when Mars is due to make a loop (April 2014, May 2016, July 2018, October 2020 and December 2022 are the next oppositions).

Jupiter

Jupiter, the father of the gods, has a majestic white light. It is as bright as Sirius, the brightest star. Like Mars, Jupiter makes a loop around the time of opposition to the Sun. Its loops are much flatter, as Jupiter's path does not deviate much from the ecliptic, and they are smaller — about 10° compared to Mars' 15° loops. The looping movement of Jupiter is not as spectacular as Mars' and also the change in brightness is not so great.

At the time of conjunction when the Sun overtakes Jupiter, it remains invisible for only 2 months, and then is visible again for 11 months. Jupiter takes 12 years to complete a revolution through the zodiac. Jupiter appears to take four steps forward and one step back, and its loops are evenly spaced around the zodiac. It is as if Jupiter likes order (Figure 100).

Saturn

In Greek mythology Saturn is Cronus, the father of Zeus or Jupiter. Saturn is not nearly as bright as Jupiter, but its light is steadier without much change in brightness. Saturn takes 30 years to complete a revolution through the zodiac, so that on average it is in one constellation for 2½ years.

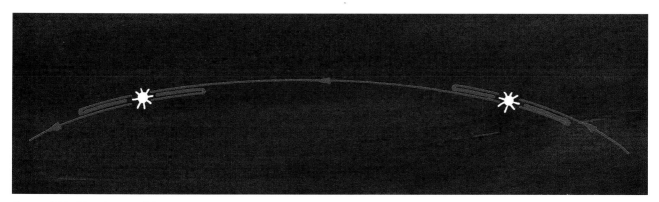

Figure 100. Two loops of Jupiter.

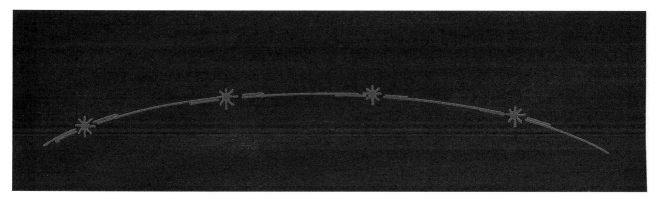

Figure 101. Loops of Saturn.

Saturn is slow, and gives the impression of being an old man. Like Jupiter it makes a loop every thirteen months, but they are smaller and flatter than Jupiter's: only 6° and close to each other. The time of retrograde is almost 5 months, and Saturn appears to take three steps forward and one step back (Figure 101).

Venus

We will describe the movement of the inferior planets, Venus and Mercury, with reference to the Sun rather than the stars. This is because they are never far from the Sun, and when they are visible at dawn or dusk the sky is often not dark enough to see many stars.

Venus is the goddess of love. Venus' appearance in the sky is quite different from that of the superior planets. Venus is always close to the Sun and can never be in opposition. There are times when Venus shines as the brightest star in the western evening sky, and at other times in the dawn sky in the east. If it is totally dark it is even possible to see a shadow cast by Venus on snow. If you know exactly where to look Venus can be seen during the day.

Figure 102. Venus as evening star. This view is as seen in the northern hemisphere. In the southern hemisphere, Venus is to the right of sunset.

Figure 103. Venus as morning star, as seen in the northern hemisphere. In the southern hemisphere, Venus is to the left of sunrise.

Venus alternates between being the evening star and the morning star. It begins its evening visibility slowly, appearing soon after sunset and noticeably less bright than at other times. Over the following weeks and months Venus increases its distance to the Sun, thus appearing for longer after sunset, and slowly becoming brighter. After about five months Venus reaches its greatest distance from the Sun. This is called the greatest eastern *elongation* (eastern, as Venus is east of the Sun). This is the time Venus is visible for longest after sunset — up to four hours, depending on latitude (Figure 102).

Having slowly reached the greatest eastern elongation, Venus now fairly rapidly reduces the distance to the Sun, though for another month its brightness increases, until it reaches its maximum brilliance. Less than a month after this Venus disappears in the *evening twilight*, ending its 7-month long *evening* visibility.

Then, after only two to three weeks of not being visible while it passes the Sun (conjunction), Venus appears again in the dawn twilight in the east. A month later it already has its greatest brilliance, all the time increasing the elongation, reaching a maximum western elongation a month after greatest brilliance (Figure 103).

Now Venus is visible for another five months, its elongation and brightness gradually decreasing, until it disappears in the dawn twilight. This ends its 7-month morning visibility. This is followed by a much longer time — about 3 months — of not being visible.

Venus swings around the Sun showing a symmetry between evening and morning visibility. The quick transition from evening to morning star is enclosed between the times of maximum brilliance and greatest elongation. Within the course of five months it passes through greatest eastern elongation (evening visibility), maximum brilliance, invisible conjunction, maximum brilliance as morning star, and greatest western elongation. The comparatively slow transition from morning to evening star is enclosed between the slow decline in elongation and visibility and after conjunction their slow increase again.

The greatest possible elongation of Venus is 47° east (evening visibility) and west (morning). Venus moves back and forth in this segment of 94° (2 x 47°) on either side of the Sun. If we remember that the Sun is moving through the zodiac in this time, Venus is like a dog on a long leash whose master walks steadily along while the dog sometimes runs ahead (eastwards) and at other times lingers behind (westwards). We can see why the conjunction from west to east (morning to evening star) takes much longer, as Venus has to catch up with its master. The transition from evening to morning star is much faster as Venus is running back in the opposite direction to the Sun.

Like all the planets, Venus' orbit is inclined to the ecliptic, so it is sometimes above and sometimes below the Sun's path. Seen from the Earth this can vary by 10° north or south. Because the angle of the ecliptic also varies through the year, Venus sometimes appears high above the horizon, and at other times much lower. That means that Venus' height above the horizon, and thus the time of visibility, is always changing. Venus appears particularly high as the evening star in spring time, or as the morning star in autumn. When this happens, the time that Venus is invisible during conjunction becomes very short. In some years it can even happen that Venus can be seen as evening and morning star on the same day in higher latitudes.

In the Arctic, Venus is visible as the evening star from September to November or as the morning star from February to March, as it is sufficiently bright to be seen in twilight or even during the day.

For Venus' visibility it is more important to know its position in relation to the Sun than to the stars. Figures 104–6 show some different paths. The path varies from year to year, both in elongation and in height above and below the ecliptic. As Venus moves from one side of the Sun to the other, it makes loops against the background of the stars. However, in temperate latitudes they are not so obvious, as fixed stars cannot be seen easily in the twilight. In tropical regions where twilight is short, the stars can be seen earlier, and so the loops can be followed more readily.

Normally during a conjunction, Venus passes the Sun above or below, like the Moon does.

The phases of Venus

With powerful binoculars or a small telescope another aspect of Venus can be seen. Venus has phases like the Moon, so has no light of its own, but reflects sunlight. However, Venus' brightness is affected by how close it is to the Earth, and at maximum brilliance Venus is actually in crescent phase.

However, if the conjunction happens when Venus is crossing the ecliptic (at a node), there is a kind of eclipse. But as Venus is so small (or so far away) there is no darkening. If the image of the Sun is projected through a telescope onto a screen you can see a small dark spot moving across the disc of the Sun. This is called a *transit* of Venus. Such transits are very rare — they only occur sixteen times in a millennium, and come in pairs 8 years apart. The last were in 2004 and 2012. The next ones will be in 2117 and 2125.

Mercury

Mercury is the messenger of the gods. It has a similar pattern of movements as Venus, but much faster, changing from evening to morning star and back about three times a year. Like Venus, Mercury is never far from the Sun, but its maximum elongation is only 28°, a little more than half of Venus'. That means Mercury is only visible for brief periods in twilight, and it is never as bright as Venus. The higher the latitude, the longer twilight lasts and the more difficult it is to see Mercury, as it is lower in the sky and the period of visibility is shorter. Many people in these regions have never seen Mercury. However, in the tropics it is possible to see Mercury three times a year as an evening star, and three times in the mornings.

Mercury, too has transits across the Sun; on average there are twelve every century. The next ones are in May 2016, November 2019, and then in 2032 and 2039.

The geocentric paths of the planets are explored in more detail on page 148f.

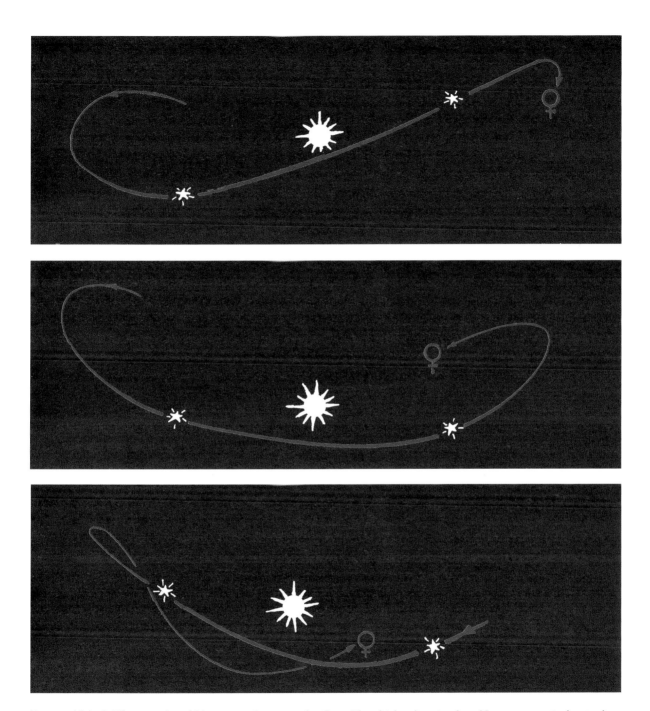

Figures 104–6. Three paths of Venus in relation to the Sun. The thicker line is when Venus passes in front of the Sun; the thinner line when it is behind the Sun. From a northern hemisphere perspective the left is evening star, the right is morning star positions. In the southern hemisphere they are the other way round.

The Planets Seen from Earth: Summary

- The planets change their position in relation to the stars.
- They remain within the constellations of the zodiac and usually move directly eastwards like the Sun and Moon.
- However, they vary their speed, and sometimes move retrograde (westwards), and they change their brightness.
- The loops of the superior planets (Mars, Jupiter, Saturn) take place at the time of opposition to the Sun, when they are also at their brightest.
- Mars' loops are the most distinct, occurring about once every two years.

- Jupiter and Saturn make a loop every year, which are smaller and flatter.
- The inferior planets, Venus and Mercury, are always close to the Sun.
- Venus shines very brightly as morning or evening star, while Mercury, being closer to the Sun, appears in twilight. Their greatest elongation is 47° and 28° respectively.
- Venus and Mercury are never in opposition to the Sun, so cannot be seen in the south (or north from southern hemisphere) at midnight.
- Transits of Venus and Mercury are rare.

15. The Planets Seen from Space

The superior planets

Until now we have simply looked at the movement of the planets as they appear in the sky. To understand the forming of the loops it is helpful to look at it from a viewpoint in space: the Sun at the centre and the Earth and a planet are moving around the Sun in a circle. For simplicity we'll assume it's a circle, though we know that all the planets move in almost circular ellipses.

The orbit of Mars is further from the Sun than the Earth's orbit. Jupiter's and Saturn's orbits are even further out. That is why they are called the superior planets, as their orbits are superior to (greater than) the Earth's. The period of their revolution increases with increasing distance: the Earth takes 1 year, Mars 2 years, Jupiter 12 years and Saturn 30 years (these are approximate values — only the Earth's is exactly 1 year). The planets' orbits are slightly inclined to the ecliptic, each one differently.

Johannes Kepler first discovered the precise mathematical relationship between the planet's period of revolution and its average distance from the Sun. (For the mathematically inclined,

Kepler's third law states that the square of the period of revolution is proportional to the cube of the distance.)

Let us look at one of the loops of Mars, as they are the most dramatic. Three things are explained by imagining the planet from space: the greatest brilliance at opposition, secondly, the fact that they move retrograde, and thirdly that the planet's distance from the ecliptic changes creating loops that face up or down, or are S- or Z-shaped.

The Earth overtakes Mars every two years, as the Earth moves twice as fast. During this time Mars, seen from the Earth, is opposite the Sun (Figure 107). Mars and Earth are as close as they will ever be, and thus Mars appears at its brightest.

To help understand this, imagine you are in a train passing a slower one. The slower one appears to be moving backwards. Something like this happens with Mars: we overtake it and it appears to move retrograde (backwards) in relation to the far distant stars. Figure 108 shows the direction in which we see Mars from Earth (blue arrows). From position 1 to position 3,

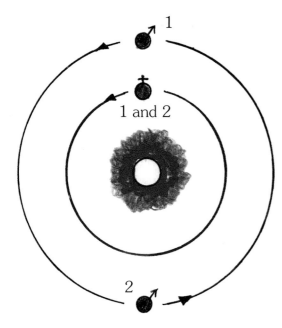

Figure 107. The orbits of Earth and Mars around the Sun. Mars takes twice as long. Position 1 is the least distance between the planets; Mars and Sun are in opposition. Position 2 is the greatest distance between the planets, and Mars is in conjunction with the Sun.

Mars appears to move retrograde. Just as with the train: from our faster moving situation, the slower one seems to move backwards.

Next we can see why the loops move above and below the ecliptic. In the previous two figures we have shown Mars as if it is in the same plane as the Earth's orbit (the ecliptic). However, this is not the case. Mars' orbit is inclined by almost 2° to the Earth's. This explains why Mars is sometimes above and sometimes below the ecliptic. These 2° are not as much as the inclination of the Moon's orbit, but when Mars is closest to the Earth, that is, in opposition to the Sun, the angle is much greater seen from the Earth (Figure 109).

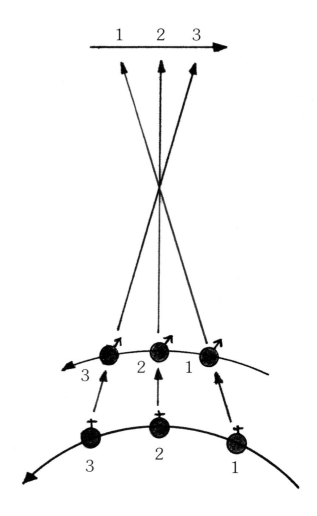

Figure 108. The Earth overtaking Mars. The blue lines are the direction of in which we see Mars from Earth. Mars appears to move backwards against the background of the stars.

Figure 109. The Earth's orbit, the ecliptic (brown), and Mars' orbit (red) which is inclined by 2°. When seen from the Earth this angle appears to be much greater.

Saturn is the furthest planet visible with the unaided eye. Its distance is again significantly further than Jupiter's. The smaller difference between apogee and perigee reduces the difference in brilliance, and the increased distance further reduces the length of its loops. While Saturn's orbit is inclined to the ecliptic by about 2.5°, its slow revolution means that there will not be much change in the height of a loop during the five months it takes for this.

Thus what would otherwise appear as a simple back-and-forth along the same track becomes a loop. To help visualise the looping in three dimensions, the lines in Figures 96–99 are thickened when Mars is closer to the Earth.

This explains the three key things about Mars' loops: greatest brilliance, retrograde motion, and the shape of the loops.

Mars' irregularities are largely owing to its elliptical orbit being less circular than that of most other planets, and thus the distances as well as speeds vary more.

Jupiter and Saturn move in a similar way to Mars. However, Jupiter's orbit is significantly further out than Mars'. This means that its brilliance does not vary as much as Mars', for the difference in distance between apogee and perigee (furthest from and closest to Earth) is relatively small. Note the difference in scale between Figures 107 and 110. Jupiter's greater distance also explains, purely from the perspective view, why Jupiter's loops are smaller and flatter. So the phenomena are similar to Mars' but not as pronounced.

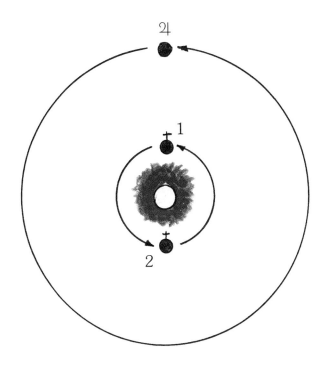

Figure 110. The Earth's orbit (brown), and Jupiter's orbit (red). At the time when Jupiter is closest to the Earth (perigee) it is in opposition to the Sun (position 1). When at apogee, Jupiter is in conjunction with the Sun. The difference in distance between perigee and apogee is not as great as with Mars.

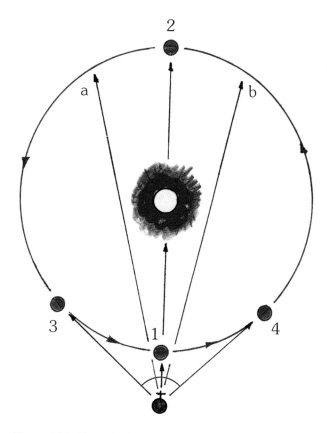

Figure 111. Venus' orbit around the Sun.
1. Inferior conjunction
2. Superior conjunction
3. Greatest eastern elongation
4. Greatest western elongation. Blue lines *a* and *b* show angle of glare

The inferior planets

Seen from space, Venus' orbit is almost circular, and smaller than the Earth's. Thus its period of revolution is shorter — it takes 225 days, roughly two thirds of a year. Figure 111 shows that it is impossible for Venus to be in opposition to the Sun. The greatest elongation (the distance from the Sun as seen from the

Earth) are the tangent lines to the orbit of Venus (positions 3 and 4). They are 47° to either side of the Sun.

We can clearly see why the length of time of Venus' invisibility varies so much. The change from evening star to morning star (the *inferior conjunction*) happens much more quickly, as Venus' path through the angle of glare is very short then. While at *superior conjunction* the change is slow, as Venus has to travel a much greater distance through the angle of glare.

Another question concerns Venus' brightness. Venus is lit by the Sun and shows phases like the Moon (visible with a small telescope). As Venus appears much bigger at apogee than at perigee, the time of maximum brilliance will be when Venus is big, but at a phase when the crescent is not too thin. This is roughly halfway between greatest elongation and inferior conjunction.

The fact that Venus only rarely transits (crosses in front of) the Sun is due to its orbit being inclined by 3.5° to the ecliptic. Seen from the Earth, this can be up to 10° above or below the ecliptic. (How this arises was explained with reference to Mars in Figure 109.)

Mercury is the closest planet to the Sun. Its revolution is only 88 days — less than three months. It cannot come as close to the Earth as Venus, and its greatest elongation is only 28°. Its orbit is inclined to the ecliptic by 7°, a relatively large inclination. It has the most pronounced ellipse of all the planets.

The Planets Seen from Space: Summary

- The planets orbit around the Sun with Mercury closest, followed by Venus, Earth, Mars, Jupiter and Saturn.
- The Earth is between the superior (further away) and the inferior (closer) planets.
- The Earth is seen like any other planet and is orbited by its satellite, the Moon.
- All the orbits are almost circular ellipses with the Sun at one of the two foci. The further from the Sun, the slower the revolution around it.
- All the planets move in the same direction around the Sun.
- Their orbits are roughly in the plane of the ecliptic (the Earth's orbit), but vary a few degrees, and thus each orbit has an ascending and a descending node.
- All planets are lit by the Sun.
- The superior planets' apparent loops (that is, as they appear to us on Earth) is explained from the perspective of being 'overtaken' by the Earth.
- The inferior planets' proximity of the Sun is the result of their orbits being within the Earth's.
- The stars are very much further from the Sun than the planets. The stars around the ecliptic form the constellations of the zodiac.

16. The Planetary Orbits

The distances of the planets

Very early on astronomers deduced something of the distances of the planets from the length of their retrograde motion and the period of their orbits. Kepler formulated this precisely in his third law (page 139).

The true distances (in kilometres or miles) were unknown, but the relative distances could be calculated. The mean (average) distance of Sun–Earth was used as an Astronomical Unit (AU). (How that actual distance was later ascertained to be 150 000 000 km is a complex story that goes beyond the scope of this book.)

The ratio of distances has intrigued astronomers for ages. Ignoring the planets discovered later with telescopes, the classic five planets, together with the Earth, move in six orbits around the Sun; there are five spaces between the orbits. Johannes Kepler was convinced of the harmony of the heavens and found a pattern that related the planetary orbits to the five Platonic solids.

The Platonic or regular solids consist of regular triangles, squares or pentagons with each point having the same number of lines and planes meeting in it (Figure 113). They are

tetrahedron (three-sided pyramid), octahedron (double pyramid), icosahedron (consisting of 20 triangles), cube and dodecahedron (consisting of 12 pentagons). Geometrically, each of these has a sphere inside that just touches the centre of each plane, and another sphere around it that just touches each of the points. Mathematicians call these the *inscribed* and the *circumscribed spheres*. For each of the Platonic solids, these spheres are in a precise ratio to each other.

Kepler arranged the planets in such a way that the circumscribed sphere of one planet was the inscribed sphere of the next, and he put them in order: Mercury – octahedron – Venus – icosahedron – Earth – dodecahedron – Mars – tetrahedron – Jupiter – cube – Saturn. He published this in his work *Mysterium Cosmographicum* in 1596. If we were to take the trouble to calculate these ratios and check them against the mean distances from the Sun, we would be surprised at the accuracy of this series. There is an engraving (Figure 114) in Kepler's book that illustrates this, and there is also a model of it in the Kepler Museum in his birthplace of Weil der Stadt in Germany.

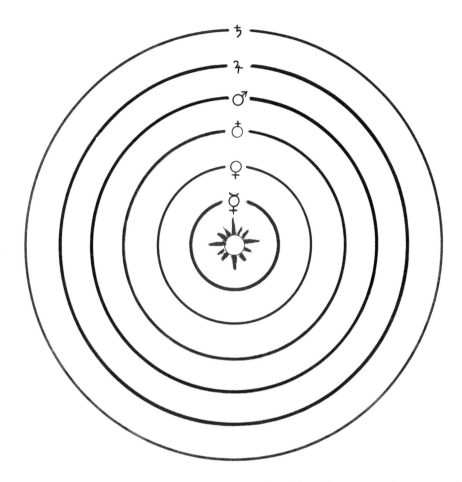

Figure 112. The heliocentric (Copernican) planet orbits around the Sun. The orbits are not drawn to scale.

octahedron ☿   icosahedron ♁   dodecahedron ♂   tetrahedron ♃   cube ♄

Figure 113. The five regular or Platonic solids. According to Kepler they determined the ratios of the planetary orbits.

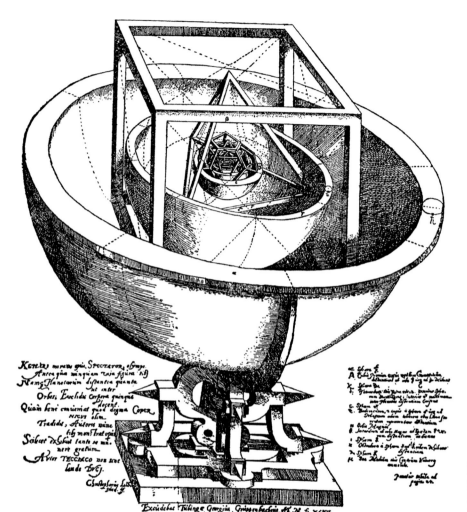

Figure 114. Kepler's Platonic solid model of the planets' orbits (from *Mysterium Cosmographicum*, 1596).

Over a century after Kepler, the astronomer Johann Bode (1747–1826) formulated a mathematical series that has since been known as Bode law (or Titius-Bode law) where after the initial two distances, each additional part is double the previous. The distances are expressed in Astronomical Units (AU).

Mercury	0.4	0.4
Venus	0.4 + (0.3 x 1)	0.7
Earth	0.4 + (0.3 x 2)	1.0
Mars	0.4 + (0.3 x 4)	1.6
???	0.4 + (0.3 x 8)	2.8
Jupiter	0.4 + (0.3 x 16)	5.2
Saturn	0.4 + (0.3 x 32)	10.0

The value between Mars and Jupiter, 2.8 is of particular interest. There is no planet visible to the unaided eye in this location. However, on the first night of 1801 the Italian astronomer Giuseppe Piazzi (1746–1826) discovered a celestial body that he named Ceres. In subsequent years a number of other such bodies were discovered in this orbit, and they were later classed as minor planets, asteroids or planetoids. Some 5 000 planetoids move in what is called the *asteroid belt* between Mars' and Jupiter's orbits. (Planetoid B612 is known as the home of Antoine de Saint-Exupéry's *Little Prince*). So in place of the three question marks in the table on the left we could put 'planetoids'.

Both Kepler's ratios and Bode's series are remarkable, though neither is quite accurate. At that time the true distances were still unknown. Once the actual mean distance of Sun–Earth (1 AU) was found, the size of the Sun could be calculated from its apparent diameter (an angle of ½°). The Sun was found to be 1 400 000 km in diameter. By comparison the Earth with a circumference of 40 000 km has a diameter of 12 700 km. So the Sun is 110 times bigger than the Earth. To get an idea of the size, imagine the Sun as a cartwheel (1.4 m in diameter) and the Earth as a hazelnut (1.3 cm diameter) about 150 metres away. Using the same scale the Moon would be a pea 38 cm from the hazelnut-Earth. Mercury would be 60 m away, and Venus 105 m, Mars 225 m, Jupiter 750 m and Saturn 1425 m away.

	Mercury	Venus	Earth	Mars	Jupiter	Saturn
Mean distance from Sun (AU)	0.4	0.7	1.0	1.5	5.2	9.5
Light time from Sun (minutes)[1]	2.3	6	8.3	13	43	80
Sidereal revolution	88^d	225^d	365^d	2^y	12^y	30^y
Mean orbit speed (km/s)	48	35	30	24	13	10
Eccentricity or orbit[2]	0.206	0.007	0.017	0.093	0.048	0.056
Inclination of orbit	7.0°	3.4°	0	1.9°	1.3°	2.5°
Rotation (days)[3]	59	–243	1	1	0.4	0.4
Maximum apogee (AU)	1.5	1.7		2.7	6.5	11
Minimum perigee (AU)	0.5	0.3		0.5	3.9	8

[1] Light time refers to the time light takes to travel from Sun to planet (for comparison light from Earth to Moon and back takes 2.4 seconds).

[2] Eccentricity is the ratio of the distance of one focus of the ellipse from the centre to its semi-major axis (half the longest diameter). A circle's eccentricity is 0; the greater the number the flatter the ellipse. Note Mercury's great eccentricity.

[3] Venus has a very slow rotation in the opposite direction to all the other planets.

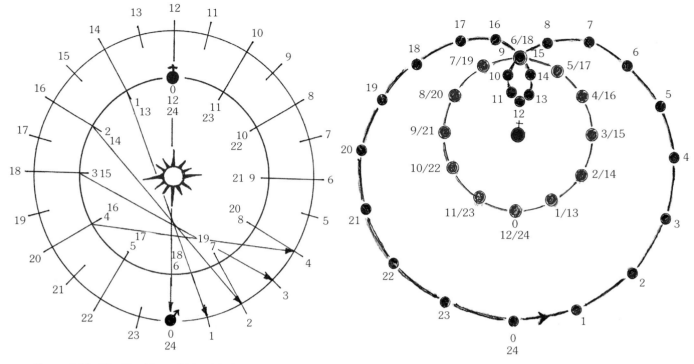

Figure 115. The Earth's and Mars' heliocentric orbits.

Figure 116. Mars' and the Sun's geocentric paths.

Geocentric paths of the superior planets

The Copernican system has the Sun in the centre of the planetary orbits. This is called *heliocentric* (from the Greek *helios*, Sun). If the Earth is viewed in the centre, and the other celestial bodies move around it, it is called *geocentric*. From our point of view (as we live on the Earth) it is interesting to know how the planets move in relation to the Earth. Let us look at the case of Mars.

First draw the orbits of Earth and Mars around the Sun, that is heliocentrically, as in Figure 115. Using a large sheet of paper, to one side draw two circles of 4 cm (1½ in) and 6 cm

(2¼ in) radius. This corresponds to their relative size, and we shall ignore the slight eccentricity of their elliptical orbit. Divide both circles into 12, and subdivide the outer circle again to make 24 divisions. These divisions show the positions of Earth and Mars at monthly intervals (as Mars takes approximately 2 years for a revolution). Number the divisions from 0 to 24 anticlockwise on each circle as shown. The inside circle has two numbers on each division as the Earth makes two revolutions during Mars' one.

Having drawn the heliocentric orbits, we now turn to the geocentric ones. On the other half

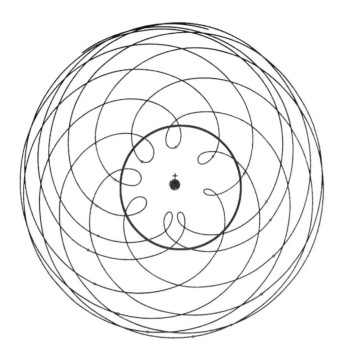

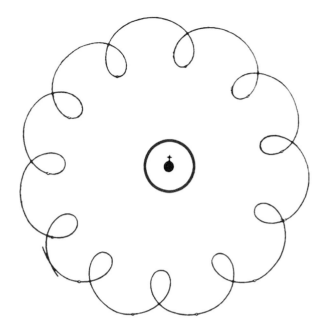

Figure 117. Geocentric path of Mars over 17 years [based on Schultz]

Figure 118. Geocentric path of Jupiter over 12 years [based on Schultz]

of the same sheet of paper, mark the Earth as the centre. Then with a pair of compasses measure the distance of Earth 0 and Mars 0 on the heliocentric drawing and transfer it to the new drawing in the same direction (in this case vertically below). We have started with the planets at greatest distance from each other: Mars is in opposition to the Sun, and is at apogee. Now continue measuring the distance and direction of each subsequent point (the first four directions are shown in blue in Figure 115).

If drawing this by hand, to transfer the angles it can help using a long ruler and sliding a set square along it. When all the points have been transferred, we have Mars' geocentric orbit. The Sun's orbit can be similarly transferred,

or a circle can be drawn around the Earth with the same radius. Finally connect the points freehand (Figure 116).

We might be alarmed to find the paths of Sun and Mars cross — is a collision possible? A moment's reflection will tell us that Mars always keeps its distance from the Sun and so when Mars crosses the Sun's path, the Sun is at a distance from Mars.

Mars' retrograde motion during the loop can clearly be seen. Here we have drawn it in a plan (from above), while in Figures 96–99 it is seen in elevation (from the side).

This drawing is simplified: we have ignored the inclination and eccentricity of Mars' orbit, and after 2 years Mars is not back in exactly the same position. Figure 117 shows Mars' actual

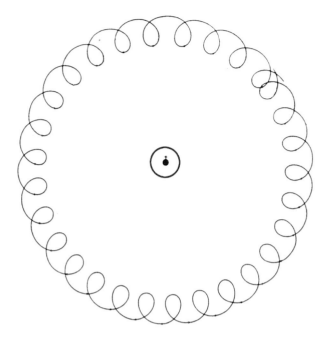

Figure 119. Geocentric path of Saturn over 30 years [based on Schultz]

Geocentric paths of the inferior planets

We can also draw Venus' geocentric path as representative of the inferior planets. Again on a large piece of paper first draw the heliocentric orbits, this time at a larger scale: 6 cm (2.5 in) radius for the Earth and 4.3 cm (1.8 in) for Venus. Divide both circles into 12, and divide each segment of the inner circle into 3 to make 36 divisions in all (Figure 120). Number the divisions on the outside circle 0 to 18 anti-clockwise from the top (some of the figures will overlap). To number the inside circle begin with 0 at the foot and count 5 divisions anticlockwise each time for subsequent numbers to 18.

Next, mark a new centre on the same sheet of paper for the Earth, and then as before for each position transfer the distance with a pair of compasses in the same direction (vertically below for position 0). Repeat for position 1 and subsequent positions to plot the monthly path of Venus around the Earth.

Connect the 18 points freehand and you have a diagram of a loop of Venus. Complete by transferring the Sun's orbit (Figure 121). We can now read the evening and morning positions of Venus from this drawing.

After eight years Venus' path repeats almost seamlessly. In this time Venus completes five loops in an elegant symmetry reminiscent of five-petalled flowers (Figure 122). There is a beauty and harmony in this pattern.

Mercury's path can be similarly drawn (Figure 123). This too is over a period of 8 years. Mercury's loops are not as even and symmetrical as Venus', giving a more unsettled impression.

orbit over 17 years (of course the inclination is also ignored in this diagram).

Similar diagrams can be made for Jupiter and Saturn over the course of 12 or 30 years (Figures 118 and 119). Note how small the orbit of the Sun is. In these figures we can recognise something of the nature of each planet: Mars' irregular dynamic, Jupiter's peace and harmony, Saturn's careful repetition.

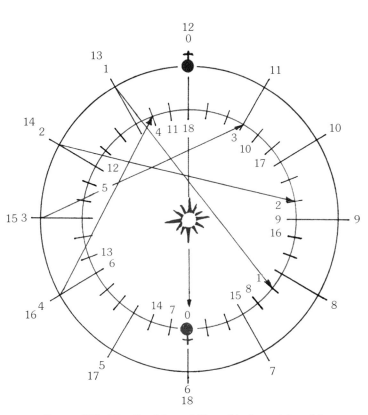

Figure 120. The Earth's and Venus' heliocentric orbits.

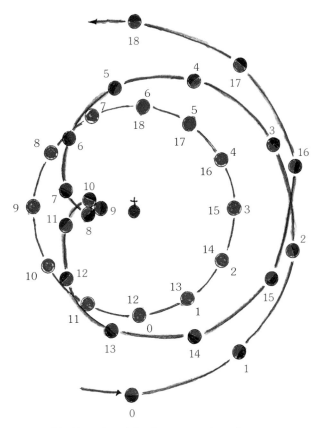

Figure 121. Venus' and Sun's geocentric paths.

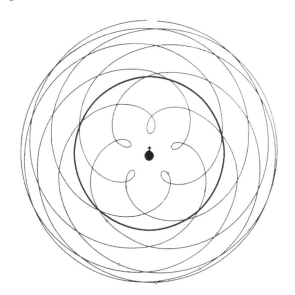

Figure 122. Venus' geocentric path almost closes after 8 years, having formed 5 loops [based on Schultz]

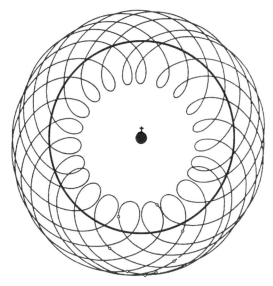

Figure 123. Mercury's geocentric path over 8 years [based on Schultz]

17. Planetary Configurations

The Sun, Moon and planets all move through the zodiac from west to east. As they move at different speeds, there are occasions when they overtake each other. These are *conjunctions*. The Moon also passes each planet every month. If for instance, the new crescent Moon is visible in the evening sky at a time when Venus is evening star, there will be a conjunction of these two (Figure 124). The Moon may pass Venus higher or lower, or even occasionally directly in front of Venus. This kind of 'eclipse' of Venus by the Moon is called an *occultation*. Because the Moon is relatively close to the Earth, an occultation will not be visible from all places on the Earth.

There are also conjunctions between planets, and these vary in appearance from passing distantly (if one planet is far below the ecliptic and another above it) or closely. Figure 125 shows a very close conjunction between Venus and Jupiter.

In principle every planet can be in conjunction with every other one. The meeting of the two slowest planets (that we can see with the unaided eye), Jupiter and Saturn, only happens once every twenty years, and is called a *Great Conjunction*. The next such conjunctions will be in December 2020 and November 2040. If the two planets are in opposition to the Sun they will be retrograde, and will then meet three times; this called a *triple conjunction*. Triple conjunctions of Jupiter and Saturn are rare — there will not be one in this century or the next. It last occurred in 1981 and will not happen again until 2238.

When two bodies are opposite each other, it is called an *opposition*. (A full Moon is the opposition of the Sun and Moon.) The superior planets are in opposition to the Sun while retrograde. There can also be oppositions between planets, which means one planet rises just as the other sets.

If two planets are at right angles to each other, it is called a *quadrature*. As one planet culminates the other is rising or setting. The Moon's first and last quarter phases are quadratures between the Moon and Sun.

Two other configurations sometimes marked are a *trine* — when the planets are 120° to each other — and a *sextile* — when they are 60° to each other.

To follow where the planets are, you need an ephemeris or almanac. Some newspapers carry a monthly night sky column, and there

Figure 124. A conjunction of the new crescent Moon and Venus in the evening (northern hemisphere view)

are websites that have similar information. *The Stargazers Almanac* carries information and is good for latitudes of Britain, southern Canada and northern United States.

It can be difficult to find the constellations of the zodiac if you are not familiar with them, but watching where Moon and planets pass can be a help to familiarise yourself with the zodiac at different times of year. Seeing where the Sun is after sunset or before sunrise can also help. Figure 126 is an example.

Planetary associations

In the descriptions so far, we have frequently had to use the words 'approximately' or 'on average'. The Sun moves at a steady pace, but not exactly

---

Astrological aspects

In astrology these planetary configurations are called *aspects*, and following symbols are used:

   ♂   conjunction
   ☍   opposition
   □   quadrature
   △   trine
   ⚹   sextile

Figure 125. Conjunction of Venus and Jupiter of February 23, 1999, seen from Prague, Czech Republic. Mercury is also visible (just above the cloud over the bright lamp in the middle). [Martin Rietze]

at the same speed. The Moon and planets move through the zodiac, but not exactly on its central line, the ecliptic. The planet's orbits are almost circles, but are slightly flattened. It is these small deviations from the regular and symmetrical that give life to the celestial phenomena. Without these irregularities, the planets would move evenly like the parts of a watch.

Since ancient times the five planets that we can see with the unaided eye, together with Sun and Moon, were seen as the classical seven planets, and were related to the days of the week, to metals, to colours, trees, grains and organs. These were not random connections but showed an insight into the nature and character of each planet and its counterpart. The table below summarises these associations.

Figure 126. The dawn sky in August 2001 seen from about latitude 50°N. From top right downwards: Saturn, Moon, Jupiter, Venus and Sun (below horizon) show the position of the zodiac. The stars in the background are Gemini. The Moon was next to Jupiter a day later, and next to Venus on the day after that. [Martin Rietze]

planet		day	metal	colour	tree	grain	organ
Sun	☉	Sunday	gold	white	ash	wheat	heart
Moon	☾	Monday	silver	lilac	cherry	rice	brain
Mars	♂	Tuesday	iron	red	oak	barley	gall
Mercury	☿	Wednesday	mercury	yellow	elm	millet	lungs
Jupiter	♃	Thursday	tin	orange	maple	rye	liver
Venus	♀	Friday	copper	green	birch	oats	kidneys
Saturn	♄	Saturday	lead	blue	beech	maize	spleen

18. Comets and Meteors

Comets

Most comets appear unexpectedly. At first they are only visible with large telescopes. They slowly become brighter and larger, until some of them can be seen by the unaided eye displaying a long tail. The tail always points away from the Sun but can reach a considerable length (Figure 127). There are reports of tails stretching right across the sky. Occasionally a comet has a second tail pointing towards the Sun. After its brightest appearance, a comet slowly loses brightness and then disappears.

A comet is a strange appearance. The name means 'long-haired' (from the Greek *kométes*). They are similar to planets in that they move in relation to the 'fixed' stars. However, they are not restricted to the zodiac.

All over the world amateur astronomers search the sky for lights that are moving in

Figure 127. Engraving by Matthäus Merian of a comet over Heidelberg, Germany in 1618. [Deutsches Museum, Munich]

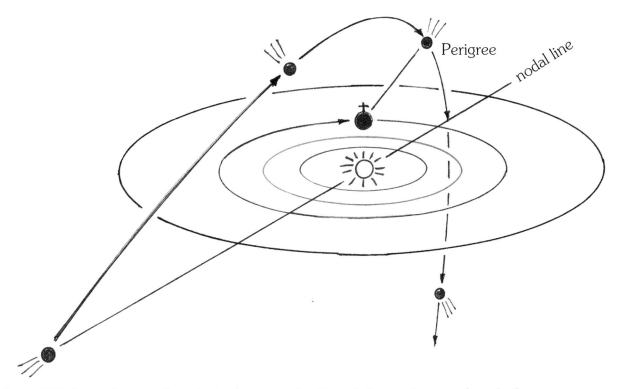

Figure 128. Comets have an elliptical orbit that is very flat. The tail always points away from the Sun.

relation to the others around them. When a new comet is discovered, it is named after the person who first reported it.

Some comets return regularly. For instance Halley's comet appears about *every* 75 years, and has been recorded since 240 BC. Its last appearance in 1986 was not as spectacular as earlier appearances. It is due again in 2061.

Comets also follow Kepler's planetary laws, though their ellipses are very flat and are sometimes parabolas (in which case they do not return). When a comet is closest to the Sun it reaches its greatest speed (Figure 128). Like the Moon and planets, it has no light of its own. Most comets' orbits take them so far away from

Sun and Earth that even with the most powerful telescopes they cannot be seen for most of their orbit. Most long-period comets are thought to originate in the Oort Cloud, a region extending about halfway to the closest stars.

Meteors

At any time when we are watching the starry sky, we might see a meteor suddenly shooting across the sky (hence their popular name of shooting or falling stars). If several fall within a short time it is called a meteor shower.

At certain times of year there is a greater

probability of seeing meteors. There are certain meteor showers which are visible every year at the same time. In the northern hemisphere one of the best know are the Perseids between July 20 and August 20, peaking around August 12 when there can be up to seventy meteors in an hour, particularly after midnight. They are named after the constellation of Perseus because the appear to come from there. The point from which they appear to come is called the *radiant point*. The best time to see them is between midnight and dawn at moonless times — new Moon or after moonset.

Other meteor showers are the Aquarids (from Aquarius), peaking around May 6 (in the southern hemisphere you can see two or three times as many meteors from this shower as in the northern hemisphere); the Orionids around October 21, and the Leonids around November 17.

Some meteor showers (like the Leonids) occasionally become a storm with thousands of meteors an hour. Such a storm occurred in 1966 over the Americas. No major storms are predicted in the next few decades (though predictions are not always accurate). Very rarely a large meteor creates a fireball.

Usually meteors burn out in the Earth's atmosphere, but sometimes debris falls on the earth. This is then called a meteorite. Most are small and create a small pit, throwing up earth. Very rarely does a large meteor strike the Earth causing a crater. Perhaps the most famous are Barringer Meteor Crater in Arizona, and Wolfe Creek Crater in Western Australia.

It is estimated that several tons of meteorites fall on the Earth daily. Most fall into the oceans (which cover almost 70% of the world's surface). Most meteorites are very small, some are no more than dust particles. But there are bigger ones, some even weighing several tons. Many museums with geological collections have meteorites. These rare finds usually consist of iron and nickel. Proof of the origin of such samples is done by slicing and polishing a sample and then etching it with acid. A fine crystalline structure of nickel-iron crystals is revealed called Widmannstätten patterns or Thomson structures (Figure 129).

The explanation for meteors is that they are small particles from comets or broken up planetoids (or even bits from the Moon or Mars) that move at great speed through the solar system. They range in size from specks of dust — the vast majority — to (rare) enormous boulders. When they enter the upper atmosphere of the Earth they heat up through friction, becoming white hot. Tiny particles

The Chelyabinsk meteor

On the morning of February 15, 2013 a huge meteor entered the Earth's atmosphere over the southern Ural region of Russia. The light from the meteor was brighter than the Sun, and eyewitnesses could feel the heat from the fireball. It exploded high in the atmosphere, generating a bright flash and shockwaves that caused widespread damage to thousands of buildings in six cities across the area.

falling off them create the characteristic trail of light.

Viewed from space, meteor showers are encountered when the Earth in its annual revolution crosses the path of a comet. Comets leave a trail of particles behind. We might think of meteors as tiny comets that have come too close to the Earth and have burnt in its atmosphere.

Just as nowadays it is almost impossible to observe a starry sky without seeing a man-made satellite moving among the stars, so when we see a meteor shower we can never be totally sure whether it is a natural one or some debris from a defunct satellite which can produce a spectacular shower of light.

Figure 129. Widmannstätten patterns or Thomson structures (enlarged) prove the material is from a meteorite. [Rupert Hochleitner]

Comets and Meteors: Summary

- Comets and meteors are phenomena that appear unexpectedly.
- Comets do not have their own light, and in this respect are similar to planets.
- Their orbit is a flattened ellipse or even parabola and is not restricted to the zodiac.
- Comets have a tail that always points away from the Sun.
- Meteors are small bits of matter hurtling through space. If they come into contact with the Earth's atmosphere the friction makes them glow white hot. So they have their own light for a brief moment.
- The trail is caused by tiny fragments breaking off and glowing.
- The remains of meteors that reach the ground are called meteorites.
- Widmannstätten patterns prove their origin in space.

Afterword

If you feel a little bewildered after reading this book, you have probably found an interest in the stars. Don't despair. It is easy to forget what has just been grasped. Keep visiting it regularly. It only takes a moment or two to look up at the stars. Try to recognise one or two constellations and find them again later, then gradually learn the surrounding ones.

Acknowledgments

I would like to thank all the people who have helped me write this book. Firstly my wife Margarethe, whose kind, critical remarks helped greatly to shape this book. Dr Peter Gmeindl, Reinhardt Schlie and Dr Helmut Rehder were colleagues who advised. Dazze Kammerl painted the pictures and Martin Rietze provided photographs. And my late friend, the astronomer Rudi Kühn must not be forgotten.

Further reading and resources

Guides and handbooks

Ridpath, Ian and Tirion, Wil, *Collins Stars and Planets: The Most Complete Guide to the Stars, Planets, Galaxies and the Solar System*, HarperCollins 2011.

Davidson, Norman, *Sky Phenomena: A Guide to Naked Eye Observations of the Heavens*, Lindisfarne Book 2004.

Schultz, Joachim, *Movement and Rhythms of the Stars*, Floris Books 2008.

Star maps and planispheres

Various planispheres are published by George Philips.

Stargazers Almanac: A Monthly Guide to the Stars and Planets, Floris Books, annual publication showing planetary position for each month.

The Maria Thun Biodynamic Calendar and *The North American Maria Thun Biodynamic Calendar* (Floris Books, annual publication) while not obviously astronomical, this calendar gives daily positions of Moon, visibility of planets and planetary configurations.

Mythology

Coder, Errol Jud, *The Constellations: Myths of the Stars*, CreateSpace 2012.

Falkner, David, *The Mythology of the Night Sky: An Amateur Astronomer's Guide to the Ancient Greek and Roman Legends*, Springer 2011.

Olcott, W.T. *Star Lore: Myths, Legends, and Facts*, 1911 (reprinted Dover 2004).

Effects of Moon

Endres, Klaus-Peter and Schad, Wolfgang, *Moon Rhythms in Nature: How Lunar Cycles Affect Living Organisms*, Floris Books 2002.

Platonic solids

Allen, Jon, *Making Geometry: Exploring Three-Dimensional Forms*, Floris Books 2012.

History

Sobel, Dava, *Longitude: The True Story of a Lone Genius who Solved the Greatest Scientific Problem of his Time*, Harper 1995; this is the story of John Harrison, inventer of the chronometer.

Models, scientific instuments, etc.

AstroMedia, www.astromediashop.co.uk or www. asttromedia.eu

Index

Explore the night skies with the
STARGAZERS' ALMANAC

A beautiful yearly guide to the night skies
designed for naked-eye astronomy
– no telescope required!

A very good, very useful Almanac
SIR PATRICK MOORE

This Almanac will show you the wonders of the night sky

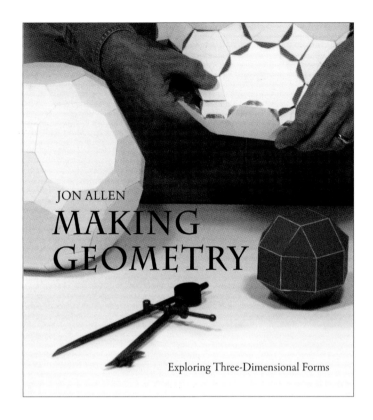

JON ALLEN

MAKING GEOMETRY

Exploring Three-Dimensional Forms

Many professionals find they need to be able to build three-dimensional shapes accurately, and understand the principles behind them. This unique book shows them how to make models of all the Platonic and Archimedean solids, as well as several other polyhedra and stellated forms.

Beginners and experienced artists and designers alike will find this book a source of practical guidance, as well as delight and inspiration which will amply repay the careful attention needed to construct the models.

The instructions and illustrations are very clear and could be used in the classroom
NETWORK REVIEW

florisbooks.co.uk

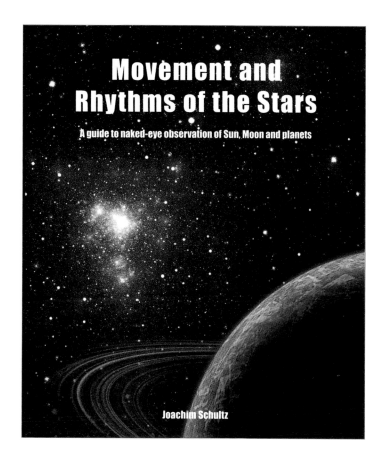

Movement and Rhythms of the Stars

A guide to naked-eye observation of Sun, Moon and planets

Joachim Schultz

This is a comprehensive guide to the basic movements we can observe in the sky. Schultz describes the daily movement of the stars from different parts of the earth (including southern hemisphere throughout). Included are the sun's pattern of the day and of the year, the moon's various periods, nodes and eclipses, as well as the planets' apparent movement and loops, conjunctions and transits.

florisbooks.co.uk

THE MARIA THUN
BIODYNAMIC CALENDAR

This useful guide shows the optimum days for sowing, pruning and harvesting various plants and crops, as well as working with bees. It includes Thun's unique insights, which go above and beyond the standard information presented in some other lunar calendars. It is presented in colour with clear symbols and explanations.

The calendar includes a pullout wallchart that can be pinned up in a barn, shed or greenhouse as a handy quick reference.

Available annually.

florisbooks.co.uk